Methods in Molecular Biology

Series Editor
John M. Walker
School of Life and Medical Sciences
University of Hertfordshire
Hatfield, Hertfordshire, UK

For further volumes:
http://www.springer.com/series/7651

For over 35 years, biological scientists have come to rely on the research protocols and methodologies in the critically acclaimed *Methods in Molecular Biology* series. The series was the first to introduce the step-by-step protocols approach that has become the standard in all biomedical protocol publishing. Each protocol is provided in readily-reproducible step-by step fashion, opening with an introductory overview, a list of the materials and reagents needed to complete the experiment, and followed by a detailed procedure that is supported with a helpful notes section offering tips and tricks of the trade as well as troubleshooting advice. These hallmark features were introduced by series editor Dr. John Walker and constitute the key ingredient in each and every volume of the *Methods in Molecular Biology* series. Tested and trusted, comprehensive and reliable, all protocols from the series are indexed in PubMed.

Schizophrenia

Methods and Protocols

Edited by

Leyre Urigüen and Rebeca Diez-Alarcia

Department of Pharmacology, University of the Basque Country UPV/EHU, Leioa, Spain

Humana Press

Editors
Leyre Urigüen
Department of Pharmacology
University of the Basque Country
UPV/EHU
Leioa, Spain

Rebeca Diez-Alarcia
Department of Pharmacology
University of the Basque Country
UPV/EHU
Leioa, Spain

ISSN 1064-3745 ISSN 1940-6029 (electronic)
Methods in Molecular Biology
ISBN 978-1-0716-3306-9 ISBN 978-1-0716-3307-6 (eBook)
https://doi.org/10.1007/978-1-0716-3307-6

This Humana imprint is published by the registered company Springer Science+Business Media, LLC, part of Springer Nature.
The registered company address is: 1 New York Plaza, New York, NY 10004, U.S.A.

Preface

Schizophrenia is a common complex disorder characterized by psychosis, cognitive dysfunction, and negative symptoms, whose etiology involves interactions between both genetic and environmental vulnerability factors.

After several decades of research, neurobiology of schizophrenia is still unknown, and no biomarkers of trace or state have yet been found. Dysfunction of dopaminergic neurotransmission was first described to contribute to the genesis of psychotic symptoms, but evidence also points to a widespread and variable involvement of other brain areas, circuits, and neurotransmission systems. Disturbances of synaptic function and abnormalities of neuronal connectivity have also been described, but the precise nature, location, and timing of these events are uncertain.

On the other hand, currently there is no cure. At present, treatment mainly consists of antipsychotic drugs combined with psychological therapies, social support, and rehabilitation, but a pressing need for more effective treatments exists.

Ongoing research attempts to elucidate the nature of vulnerability, environmental risk factors, genetic and neurobiological mechanisms, and clinical manifestations of schizophrenia. Advances in research methods and development of new approaches will lead to a great progress in understanding the disorder, and the developing of new therapeutic strategies.

The aim of this book is to compile different methods currently applied in the research of schizophrenia, from the most molecular techniques to in vivo approaches both in human beings and in animal models of the disease.

Leioa, Spain

Leyre Urigüen
Rebeca Diez-Alarcia

Contents

Contributors

OIER AIZPURUA-OLAIZOLA • *Sovereign Fields S.L., 20006, San Sebastian, Basque Country, Spain*

ROSA M. BEÑO-RUIZ-DE-LA-SIERRA • *Psychiatry Department, School of Medicine, University of Valladolid, Valladolid, Spain*

MICHEL BOUVIER • *Department of Biochemistry and Molecular Medicine, Institute for Research in Immunology and Cancer (IRIC), Université de Montréal, Montréal, QC, Canada*

JANIRE CABEZAS-GARDUÑO • *Early Stages of Psychosis, Biocruces Bizkaia Health Research Institute, Barakaldo, Spain; Department of Psychiatry, Cruces University Hospital, Barakaldo, Spain*

BRIAN DEAN • *The Synapse Biology and Cognition Laboratory, The Florey Institute for Neuroscience and Mental Health, Parkville, VIC, Australia*

R. DIEZ-ALARCIA • *Department of Pharmacology, University of the Basque Country UPV/EHU, Leioa, Bizkaia, Spain; Centro de Investigación Biomédica en Red de Salud Mental, Instituto de Salud Carlos III, Bizkaia, Spain; Biocruces Bizkaia Health Research Institute, Bizkaia, Spain*

ÁLVARO DÍEZ-REVUELTA • *Psychiatry Department, School of Medicine, University of Valladolid, Valladolid, Spain*

NESTOR ETXEBARRIA • *Department of Analytical Chemistry, Faculty of Science and Technology, University of the Basque Country (UPV/EHU), Leioa, Basque Country, Spain; Research Centre for Experimental Marine Biology and Biotechnology (PIE), University of the Basque Country (UPV/EHU), Plentzia, Basque Country, Spain*

INÉS FERNÁNDEZ-LINSENBARTH • *Psychiatry Department, School of Medicine, University of Valladolid, Valladolid, Spain*

SUPRIYA A. GAITONDE • *Department of Biochemistry and Molecular Medicine, Institute for Research in Immunology and Cancer (IRIC), Université de Montréal, Montréal, QC, Canada*

JAVIER GONZÁLEZ-MAESO • *Department of Physiology and Biophysics, Virginia Commonwealth University School of Medicine, Richmond, VA, USA*

CAROLINA GUBERT • *Florey Institute of Neuroscience and Mental Health, Melbourne Brain Centre, University of Melbourne, Parkville, VIC, Australia*

ANTHONY J. HANNAN • *Florey Institute of Neuroscience and Mental Health, Melbourne Brain Centre, University of Melbourne, Parkville, VIC, Australia; Department of Anatomy and Physiology, University of Melbourne, Parkville, VIC, Australia*

ALAINA M. JASTER • *Department of Physiology and Biophysics, Virginia Commonwealth University School of Medicine, Richmond, VA, USA; Department of Pharmacology and Toxicology, Virginia Commonwealth University School of Medicine, Richmond, VA, USA*

GERALDINE KONG • *Florey Institute of Neuroscience and Mental Health, Melbourne Brain Centre, University of Melbourne, Parkville, VIC, Australia; Peter Doherty Institute of Infection and Immunity, University of Melbourne, Parkville, VIC, Australia*

NOEMI LOSADA-LEÓN • *Early Stages of Psychosis, Biocruces Bizkaia Health Research Institute, Barakaldo, Spain; Department of Psychiatry, Cruces University Hospital, Barakaldo, Spain*

Oihane Mentxaka • *Early Stages of Psychosis, Biocruces Bizkaia Health Research Institute, Barakaldo, Spain; Department of Neurosciences, University of the Basque Country UPV/EHU, Leioa, Spain; Department of Psychiatry, Cruces University Hospital, Barakaldo, Spain*

I. Muneta-Arrate • *Department of Pharmacology, University of the Basque Country UPV/EHU, Leioa, Bizkaia, Spain; Centro de Investigación Biomédica en Red de Salud Mental, Instituto de Salud Carlos III, Bizkaia, Spain*

Estibaliz Olabarrieta • *Department of Pharmacology, Medicine and Nursing Faculty, University of the Basque Country UPV/EHU, Leioa, Spain*

Maitane Olivares • *Department of Analytical Chemistry, Faculty of Science and Technology, University of the Basque Country (UPV/EHU), Leioa, Basque Country, Spain; Research Centre for Experimental Marine Biology and Biotechnology (PIE), University of the Basque Country (UPV/EHU), Plentzia, Basque Country, Spain*

Geoffrey Pavey • *The Victorian Brain Bank Network, The Florey Institute for Neuroscience and Mental Health, Parkville, VIC, Australia*

María Recio-Barbero • *Early Stages of Psychosis, Biocruces Bizkaia Health Research Institute, Barakaldo, Spain; Department of Neurosciences, University of the Basque Country UPV/EHU, Leioa, Spain*

Alejandro Roig-Herrero • *Psychiatry Department, School of Medicine, University of Valladolid, Valladolid, Spain*

Markel San Nicolas • *Department of Analytical Chemistry, Faculty of Science and Technology, University of the Basque Country (UPV/EHU), Leioa, Basque Country, Spain; Sovereign Fields S.L., 20006, San Sebastian, Basque Country, Spain*

Rafael Segarra • *Early Stages of Psychosis, Biocruces Bizkaia Health Research Institute, Barakaldo, Spain; Department of Neurosciences, University of the Basque Country UPV/EHU, Leioa, Spain; Department of Psychiatry, Cruces University Hospital, Barakaldo, Spain; Centro de Investigación Biomédica en Red de Salud Mental CIBERSAM, Leioa, Spain*

Megan Snelleksz • *The Synapse Biology and Cognition Laboratory, The Florey Institute for Neuroscience and Mental Health, Parkville, VIC, Australia*

Lide Totorikaguena • *Department of Physiology, Medicine and Nursing Faculty, University of the Basque Country UPV/EHU, Leioa, Spain*

Paula Unzueta-Larrinaga • *Biocruces Bizkaia Health Research Institute, Barakaldo, Spain; Department of Pharmacology, University of the Basque Country UPV/EHU, Leioa, Spain*

Leyre Urigüen • *Biocruces Bizkaia Health Research Institute, Barakaldo, Spain; Department of Pharmacology, University of the Basque Country UPV/EHU, Leioa, Spain; Centro de Investigación Biomédica en Red de Salud Mental CIBERSAM, Madrid, Spain*

Aresatz Usobiaga • *Department of Analytical Chemistry, Faculty of Science and Technology, University of the Basque Country (UPV/EHU), Leioa, Basque Country, Spain; Research Centre for Experimental Marine Biology and Biotechnology (PIE), University of the Basque Country (UPV/EHU), Plentzia, Basque Country, Spain*

Aitor Villate • *Department of Analytical Chemistry, Faculty of Science and Technology, University of the Basque Country (UPV/EHU), Leioa, Basque Country, Spain*

Chapter 1

A Novel Method to Estimate Levels of Receptors Using an Allosteric Modulator: Focus on Muscarinic M1 Receptor

Geoffrey Pavey, Megan Snelleksz, and Brian Dean

Abstract

This chapter outlines some of the general principles that need to be considered when developing a radioligand binding assay to measure the affinity and density of radioligand binding to a receptor in tissue or on cells. In addition it describes an innovative step forward in using radioligand binding assays to measure levels of muscarinic M1 receptors in human postmortem CNS, using both membrane binding and in situ radioligand binding. These examples show how, using receptor-specific allosteric modulators, it is possible to gain an estimate of the density of a single receptor using a radioligand that is not totally specific to the target site of interest. Given there is a growing understanding that there are problems with antibodies not showing specificity to their supposed target protein, well-characterized radioligand binding techniques still provide an important tool when studying receptor density in tissues and cells.

Key words Radioligand binding, Autoradiography, Muscarinic receptors, Muscarinic M1 receptor, Tissue homogenization, Phosphor imaging

1 Introduction

Radioligand binding assays can be used to measure a number of parameters relating to the binding of two molecules [1], but this chapter will specifically focus on the use of radioligand binding to obtain a measure of the density of muscarinic M1 receptors in a tissue or on cells.

It has been argued that the lack of specificity of some radioligands for a single binding site is a major disadvantage in using radioligand binding to measure levels of proteins such as receptors [2]. This has led to the use of different transcriptomic technologies [3, 4] and "protein-specific" technologies such as Western blotting [5], rather than radioligand binding, to estimate the level of expression and density of receptors in different tissues. While measuring levels of receptor mRNA clearly gives a good measure of receptor expression, in many cases, levels of receptor expression do not

Leyre Urigüen and Rebeca Diez-Alarcia (eds.), *Schizophrenia: Methods and Protocols*, Methods in Molecular Biology, vol. 2687, https://doi.org/10.1007/978-1-0716-3307-6_1,

correlate with the more function-related measure of the steady-state receptor protein level [6]. In addition, and of concern, is the demonstration that many antibodies used in research lack specificity [7, 8] adding to existing concerns about the use of Western blotting [9] and enzyme-linked immunoassays [10] to give an accurate measure of protein levels in tissue. To overcome these problems, well-defined radioligand binding techniques can still be used to gain a good estimate of levels of receptors in tissue or on cells.

As in other approaches, there are a number of factors that should be considered when designing radioligand binding assays to measure levels of receptor density. First, there must be a radioligand that binds to the receptor of interest with a high affinity [11]. If there is a high-affinity radioligand available that is not totally specific to the receptor of interest, it is possible to develop relatively specific receptor binding assays using the physicochemical and pharmacological properties of the receptor. Thus, shorter incubation times will increase the specificity of a radioligand to sites where the radioligand has a fast on-rate [12]. There are a number of approaches to increasing the specificity of radioligand binding assays that are independent of the specificity of the radioligand used. Most commonly, the specificity of a radioligand binding assay for a receptor is increased by using an unlabeled drug (i.e., without a radioactive isotope) that shares a single binding site in common with the radioligand as a displacing agent [13, 14]. In addition, where a radioligand binds to two binding sites, an unlabelled drug specific to one of those binding sites can be used to block the binding of the radioligand to that site and the levels of the second site can then be measured by using an unlabelled drug specific to the site of interest to displace the remaining bound radioligand; we used such an approach to develop and assay for the dopamine transporter [15]. In addition, the sequential use of multiple displacing agents with different binding sites can allow the quantification of a number of different binding sites using the same radioligand [16]. Another significant advance is the availability of cloned proteins and tissue from a gene knockout cell or animal that can be used to better measure the specificity of a radioligand binding assay for any binding site of interest [12, 17]. Finally, another innovative use of receptor binding assays is to begin to identify endogenous factors that modulate the binding activity of sites of interest [18].

In addition to optimizing the binding of a radioligand to maximize specificity of binding to a receptor of interest, there are other experimental variables that need to be considered when developing a radioligand binding protocol. These would include using either crude particulate homogenate [19], membrane-enriched preparations [20], or tissue sections [19] to measure radioligand binding. In addition, while the gold standard approach to gaining a measure of radioligand binding sites in tissue or on cells would be to do a full binding curve [20], where the availability of tissue is limited, a good estimate of binding can be achieved using a single-point saturation binding technique [21]. Importantly, in using a single-point saturation analysis, it is critical that the concentration of radioligand used is equal to at least three times its Kd for the binding site of interest to ensure all available binding sites are occupied [22, 23]. This methodological approach is particularly critical in ensuring the accurate estimation of the density of binding sites when there are small changes in the affinity of radioligand binding associated with different disease states [24]. The final decision on which technique to use will be influenced by the abundance of tissue or cells and the funding available for the reagents needed to complete a radioligand binding study (Table 1).

There are also innovative variations to these basic radioligand binding methodologies that allow novel data to be gathered on receptor function. One important discovery influencing the development of more innovative radioligand binding assays was the discovery that many receptors have allosteric binding sites [25] that are distal to the orthosteric binding site at which neurotransmitter binding occurs. Positive allosteric modulators act to increase the affinity of the orthosteric binding site on the muscarinic receptor to which they bind [25], and hence measuring the change of affinity of the orthosteric binding site is a surrogate measure of the number of that receptor in a tissue. Important to this chapter, there are now drugs that show total specificity to the muscarinic M1 receptor [26, 27], and we have used such a drug, the positive allosteric modulator [1-(4-methoxybenzyl)-4-oxo-1,4-dihydroquinoline-3-carboxylic acid] (BQCA), to develop techniques to measure levels of the muscarinic M1 receptor in human postmortem central nervous system (CNS) [28, 29].

2 Materials

This methodology requires deionized water (dH_2O) (18 MΩ.cm at 25 °C) and analytical-grade reagents.

2.1 Solutions

1. *Gelatin solution*: solution of 0.5% gelatin and 0.05% chromium potassium sulfate.
2. *Assay buffer: 20 mM HEPES buffer.* To prepare assay buffer, on a magnetic stirrer, dissolve 4.76 g (20 mM) HEPES, 5.84 g sodium chloride (100 mM; NaCl), and 0.952 g magnesium chloride (10 mM; $MgCl_2$) into 900 mL dH_2O. Adjust to pH 7.4 with a small volume of 10 M NaOH. Make up to 1 L with dH_2O, preferably in a volumetric flask (*see* **Note 1**).
3. *1% DMSO solution*: A solution of 1% DMSO is prepared by diluting 50 μL DMSO into 5 mL assay buffer.

Table 1
Advantages and disadvantages of different approaches to using radioligand binding to measure receptor levels on tissue

Full binding curve		Single-point saturation analysis			
Advantages	**Disadvantages**	**Advantages**		**Disadvantages**	
Full binding parameters (Kd and Bmax)	Uses more tissue Uses more reagents	Gives a good estimate of density of radioligand binding sites Uses less tissue and reagents		A good measure of affinity of radioligand to the tissue is required to ensure maximal binding	
Crude particulate homogenate		**Membrane-enriched preparations**		**Tissue sections**	
Advantages	**Disadvantages**	**Advantages**	**Disadvantages**	**Advantages**	**Disadvantages**
Relatively simple to prepare	Will have higher levels of nonspecific binding Does not give spatial resolution	Will contain more receptors per unit of protein and have lower nonspecific binding and may therefore give increased binding sensitivity	Difficult and time consuming to prepare Needs access to high-speed centrifuges Does not give spatial resolution	Gives spatially resolved binding allowing binding in discrete cytoarchitectural subregions to be measured	Long incubation to either film or phosphor image plates Needs access to darkroom or phosphor imaging equipment

Table 2
Preparing the acetylcholine stock dilutions required to displace [^{3}H]NMS from cortical muscarinic receptors

Acetylcholine concentration	Acetylcholine (μL)	Buffer (μL)
10 mM	200 of 0.1 M	1800
5 mM	500 of 10 mM	500
1 mM	200 of 5 mM	1800
500 μM	500 of 1 mM	500
100 μM	200 of 500 μM	1800
50 μM	500 of 100 μM	500
10 μM	200 of 100 μM	1800
5 μM	500 of 10 μM	500
1 μM	200 of 5 μM	1800
100 nM	100 of 1 μM	900
10 nM	100 of 100 nM	900

Note: These prepared solutions are then diluted 1:10 into the assay test tubes to give a final assay concentration range of 1 nM to 1 mM

2.2 Preparation of 1 mM Acetylcholine

1. Prepare a stock solution of 0.1 M acetylcholine by dissolving 18.17 mg per 1 mL of assay buffer on the day of the assay.
2. For measuring the displacement of [^{3}H]NMS, dilute the 0.1 M solution of acetylcholine as detailed in Table 2 (*see* **Note 2**).

2.3 Preparation of Radioligand [^{3}H]NMS for Membrane-Binding Assay

1. Add 1 μL of [^{3}H]NMS (scopolamine methyl chloride, [N-methyl-^{3}H]; specific activity of 85.5 Ci/mmol (1 mCi/mL; 11.7 μM)) from the stock solution provided by the manufacture to 899.7 μL assay buffer to give a concentration of 13 nM (*see* **Note 3**).
2. Ensure the concentration of this first dilution of [^{3}H]NMS is correct by placing 25 μL of the 13 nM solution in 5 mL of liquid scintillation fluid and counting its level of radioactivity in a liquid scintillation counter. The level of radioactivity in disintegrations per minute (d.p.m) should match what would be expected from a 13 nM of [^{3}H]NMS.
3. Dilute the 13 nM of [^{3}H]NMS solution 1:10 into assay buffer to give a 1.3 nM solution of [^{3}H]NMS. This solution is again diluted 1:10 in assay buffer in the final assay to give 1.3 nM [^{3}H]NMS.

2.4 Preparation of Radioligand [^{3}H]NMS for Autoradiography

1. Add 1 μL of [^{3}H]NMS (scopolamine methyl chloride, [N-methyl-3H]; specific activity of 85.5 Ci/mmol (1 mCi/mL; 11.7 μM)) from the stock solution provided by the manufacture to 2.924 mL assay buffer to give a solution of 4 nM [^{3}H]NMS. This solution is diluted 1:10 in assay buffer in the final assay to give 0.4 nM [^{3}H]NMS.
2. Ensure the concentration of this first dilution of [^{3}H]NMS is correct by placing 25 μL of the 4 nM solution in 5 mL of liquid scintillation fluid and counting its level of radioactivity in a liquid scintillation counter. The level of radioactivity in disintegrations per minute (d.p.m) should match what would be expected from a 4 nM of [^{3}H]NMS.

2.5 Preparation of BQCA

1. Prepare 3 mM BQCA by dissolving 0.093 mg BQCA per 100 μL dimethyl sulfoxide (DMSO) (*see* **Note 4**).
2. Dilute the 3 mM BQCA 1:100 in assay buffer to a concentration of 30 μM BQCA.

2.6 Equipment

Powered homogenizers (e.g., Janke and Kunkel Ultra-Turrax Homogenizer)

Beckman J2-HS centrifuge with JA 20.1

Brandel cell harvester

FujiFilm BAS5000 phosphor imager (e.g., Leica CM1950)

Liquid scintillation counter (e.g., TRI-CARB 4910TR 110 V Liquid Scintillation Counter)

3 Methods

3.1 BQCA Acetylcholine Displacement Shift Assay: Homogenates

3.1.1 Membrane Homogenates Preparation

1. Dissect approximately 1 g of the area of postmortem CNS tissue to be studied (e.g., gray matter from the human frontal cortex).
2. Place tissue in a glass/PFTE Potter Elvehjem tissue grinder, add 5 mL of ice-cold assay buffer, and homogenize with ten strokes, by hand, of the homogenizer.
3. Transfer the resulting homogenate to a 15 mL polycarbonate centrifuge tube and centrifuge at 20,000× *g* for 20 min at 4 °C.
4. Discard the resulting supernatant and resuspend the pellet in the original volume of ice-cold assay buffer.
5. Repeat **steps 3** and **4** twice and disperse the final pellet in 1 mL of assay buffer.
6. Measure the protein level of each homogenate (*see* Subheading 3.1.2).
7. Aliquot the homogenate into appropriate-sized volumes and store it at −80 °C until required.

3.1.2 Protein Measurement

Protein levels in CNS homogenates can be measured using any recognized methodology, and we use the Bio-Rad protein assay strictly following the manufacturer's instructions. This assay is based on the method developed by Bradford [30] and involves adding an acidic dye to a solution containing proteins. Comparing the absorbance of the resulting solution at 595 nm in a spectrometer or microplate reader, to the absorbance of a standard curve prepared from solutions provided, gives an accurate measurement of the protein concentration of the solution. The chemical principles by which the dye and proteins interact are provided in the manufacturer's instructions.

3.1.3 BQCA Acetylcholine Displacement Shift Assay

The principals underlying this assay are detailed elsewhere [28]. Here we describe the fully optimized assay we developed to measure BQCA-sensitive [^{3}H]NMS binding as a measure of muscarinic M1 receptors in postmortem human CNS.

1. This assay is performed in 5 mL polystyrene test tubes in triplicate for each concentration of acetylcholine. All tubes and reagents must be kept on ice and added to as described in Table 3. *It is important to add the membrane last and vortex each tube briefly as this starts the binding of radioligand, acetylcholine, and BQCA simultaneously to the muscarinic receptors in the cortex.*
2. After a 2 h incubation at room temperature, the reaction in each tube is terminated by filtering the solution in each tube through Whatman GF/B glass filter paper soaked for 1 h in 0.1% polyethylenimine (PEI) using a Brandel cell harvester.
3. Each filter is then washed 3 × 5 mL with ice-cold assay buffer to remove unbound radioligand and transferred into scintillation tubes.
4. Add 5 mL of liquid scintillation solution to each tube.
5. Each tube is placed in a liquid scintillation counter and the radioactivity on each filter counted.

While the effect of BQCA on the displacement of [^{3}H]NMS by acetylcholine is reflected by a change in the median effective dose (ED_{50}) of the two displacement curves, this value did not correlate strongly with the level of cortical protein and hence is not a good measure of the number of cortical muscarinic M1 receptors in tissue. Instead of that, using the methodology described in this chapter, the best measure of the density of muscarinic M1 receptors is the measure of the area within the displacement curves of [^{3}H]NMS by acetylcholine in the absence and presence of BQCA [28]. To make this measurement:

- Subtract the [^{3}H]NMS radioactivity bound at each concentration of acetylcholine in the presence of BQCA from that in the absence of BQCA.

Table 3
Preparing the assay to measure the displacement of [^{3}H]NMS from human cortex by acetylcholine in the absence and presence of 3 μM BQCA (*see* Notes 1 and 3)

Tube	Assay buffer (μL)	1% DMSO (μL)	Acetylcholine (μL)	3 μM BQCA in 1% DMSO (μL)	1.3 nM [^{3}H]NMS (μL)	Membrane (μL)
1–3	700	100	0	–	100	100
4–6	600	100	100 (10 nM)	–	100	100
7–9	600	100	100 (100 nM)	–	100	100
10–12	600	100	100 (1 μM)	–	100	100
13–15	600	100	100 (5 μM)	–	100	100
16–18	600	100	100 (10 μM)	–	100	100
19–21	600	100	100 (50 μM)	–	100	100
22–24	600	100	100 (100 μM)	–	100	100
25–27	600	100	100 (500 μM)	–	100	100
28–30	600	100	100 (1 mM)	–	100	100
31–33	600	100	100 (5 mM)	–	100	100
33–36	600	100	100 (10 mM)	–	100	100
37–39	700	–	0	100	100	100
40–42	600	–	100 (10 nM)	100	100	100
43–45	600	–	100 (100 nM)	100	100	100
46–48	600	–	100 (1 μM)	100	100	100
49–51	600	–	100 (5 μM)	100	100	100
52–54	600	–	100 (10 μM)	100	100	100
55–57	600	–	100 (50 μM)	100	100	100
58–60	600	–	100 (100 μM)	100	100	100
61–63	600	–	100 (500 μM)	100	100	100
64–66	600	–	100 (1 mM)	100	100	100
67–69	600	–	100 (5 mM)	100	100	100
70–72	600	–	100 (10 mM)	100	100	100

- Enter these data into software that can calculate area under curves making the X axis ranging from 1 to 12 (the number of concentrations of acetylcholine used) and the Y axis as the binding in the absence minus that in the presence of BQCA.
- Calculate the area under the resulting curve, which is the measure of the density of muscarinic M1 receptors in the tissue.

3.2 BQCA Acetylcholine Displacement Shift Assay: In Situ Radioligand Binding with Autoradiography

The methodology described to measure BQCA-mediated changes in the displacement of [^{3}H]NMS from crude homogenate by acetylcholine requires a significant amount of tissue and reagents. Hence, to use less irreplaceable tissue, we developed a methodology using in situ radioligand binding with autoradiography that uses single-point saturation analysis to measure BQCA-mediated changes in acetylcholine displacement of [^{3}H]NMS [29]. The advantages of using this methodology are that it uses much less tissue, requires much less reagents, and gives spatial readouts that allows the measurement of binding to discrete areas within complex regions of the CNS such as the hippocampus. We have shown the results obtained with this methodology are comparable to those obtained using the membrane-binding assay [29].

3.2.1 Preparation of Frozen Tissue Sections

1. First, all microscope slides (25 mm × 76 mm) are coated in gelatin by placing the slides in racks and dipping them into warm gelatin solution (40–50 °C). The slides are then removed and drained onto paper tissue and are then covered with foil to keep off dust and air-dried overnight in a dust-free container at room temperature (*see* **Note 5**).
2. In a freezing microtome, 20 μm frozen coronal sections are cut from postmortem CNS tissue at −18 °C and stored at −80 °C until assay.
3. Six consecutive frozen sections from each case to be studied are mounted onto two gelatin-coated glass microscope slides (three sections per slide).

3.2.2 BQCA Acetylcholine Displacement Shift Autoradiographic Assay

1. On the assay day, the sections are removed from −80 °C and allow to air-dry face up until they reach room temperature (~30 min).
2. While the sections are drying, the solutions required to measure BQCA-mediated [^{3}H]NMS binding are mixed together using the volumes detailed in Table 4.
3. Once dry, the slides containing the frozen sections are placed in a moist box that can either be purchased or made in house (to prepare moist box in-house, see [31]).
4. Once the slides are in the moist box, 250 μL of solution 1 (Table 4) are carefully placed over the three sections used to measure the binding of [^{3}H]NMS in the presence of a standardized dose of acetylcholine using the "drop assay" technique

Table 4
Preparing the solutions to measure BQCA-mediated changes in acetylcholine displacement of [^{3}H]NMS using in situ radioligand binding with autoradiography

Solution	Assay buffer (μL)	0.1% DMSO (μL)	1 mM acetylcholine (μL)	3 μM BQCA in 0.1% DMSO (μL)	0.4 nM [^{3}H]NMS (μL)
1	700	100	100	–	100
2	700	–	100	100	100

that allows the solution to cover each section and to be held in place by surface tension. In addition, 250 μL of solution 2 (Table 4) are carefully placed over the three sections used to measure the effect of BQCA on the binding of [^{3}H]NMS in the presence the of a standardized dose of acetylcholine using the "drop assay" technique (*see* **Note** 7).

5. Once all the solutions have been added to all sections in a moist box, the lid should be placed on the moist box to minimize evaporation.
6. After incubating for 2 h at room temperature, all the slides are immersed in two lots of ice-cold assay buffer, sequentially, for 3 min. The sections are then briefly dipped in ice-cold deionized water for 10 s to remove the residual buffer salts and then dried rapidly under a stream of warm air from a hairdryer.
7. All sections are then partially fixed by being placed in paraformaldehyde vapor in a desiccator chamber.
8. Next, the partially fixed sections are apposed to phospho-imaging plates along with tritium standards for 7 days or until clear images are detectable when the plates are scanned.
9. After 7 days, phospho-imaging plates are scanned in a phosphor-imager. The resulting images can be used to measure the density of [^{3}H]NMS binding as d.p.m./mg estimated tissue equivalents by comparing the density of each tissue section image to a standard curve derived from standards containing a known amount of tritium radioactivity [32] using image density quantification software such as AIS imaging software.

As with membrane binding, BQCA increased the affinity of the orthosteric binding site enabling 1 mM acetylcholine to displace more [^{3}H]NMS from muscarinic M1 receptors in cortical tissue sections. Thus, BQCA-sensitive displacement of [^{3}H]NMS could be calculated by subtracting the level of [^{3}H]NMS binding in the presence of acetylcholine and BQCA (solution 2) from that in the absence of BQCA (solution 1).

In this chapter, we detail how a receptor-specific allosteric modulator can be used to gain a measure of the muscarinic M1 receptor in postmortem human CNS. It should be noted that the availability of cloned receptors and tissue from muscarinic receptor knockout mice has opened up the opportunity to better show the specificity of radioligand binding to postmortem CNS [12, 17]. Importantly, similar approaches using other receptor specific allosteric modulators [25] would lead to a new family of radioligand binding assay that would provide new insights into how receptors are expressed in postmortem CNS. Such radioligand binding assays will be important tool given some protein levels cannot be measured due to the lack of an antibody with specificity to proteins of interest and when the levels of a protein are too low to be quantified using other approaches such as mass spectrometry [33].

4 Notes

1. The assay buffer used in this methodology (20 mM HEPES, 100 mM NaCl, and 10 mM MgCl, pH 7.4) has been previously demonstrated to be optimum for use with [^{3}H]NMS, acetylcholine, and [1-(4-methoxybenzyl)-4-oxo-1,4-dihydroquinoline-3-carboxylic acid] (BQCA) [26, 27], the three active components in this methodology. For any other experimental design, preliminary buffer optimization should be carried out.
2. In our hands, this stepwise approach to preparing acetylcholine has proven more practical than preparing a 1 mM acetylcholine solution directly. As described in Table 2, dilutions are prepared 10× as they will be diluted 1:10 into the assay test tubes giving a final assay concentration range of 1 nM to 1 mM. The same apply to all the other regents for which the dilution factor in the final incubation volume must be considered.
3. Radioactive ligands can be obtained from a number of sources, and some companies will prepare customized radioligands and, if requested, provide the radioligand in a purified form. Specific activity and concentration of the radioligand may change from one batch to another. Thus, the volume of stock radioligand solution used must be reviewed for each batch and radioligand.
4. As BQCA is dissolved in DMSO, 100 μL of 1% DMSO solution should be added into those tubes not containing BQCA to give a final concentration of 0.1% DMSO in all incubation conditions.
5. The gelatin solution should only be used once and discarded.
6. In methodology using in situ radioligand binding with autoradiography, we used 1 mM acetylcholine because we had shown this was the lowest dose of acetylcholine that constantly displaced some [^{3}H]NMS from cortical sections.

7. The "drop assay" technique requires the careful application of solutions 1 and 2 to the frozen sections so that surface tension retains the solution on each section and prevents the solution from "spilling" onto the glass slide. If this proves problematic, a circle can be drawn round each section with a chinagraph pencil to help restrict the spread of solutions 1 and 2.

References

1. McKinney M (2006) Practical aspects of radioligand binding. Curr Protoc Pharmacol Suppl 33:1.3.1–1.3.42
2. McCrea KE, Herzog H (2000) Radioligand binding studies. In: Balasubramania A (ed) Neuropeptide Y protocols. Humana Press, London
3. Mancama D, Arranz MJ, Landau S, Kerwin R (2003) Reduced expression of the muscarinic 1 receptor cortical subtype in schizophrenia. Am J Med Genet B Neuropsychiatr Genet 119:2–6
4. Dean B, McLeod M, Keriakous D, McKenzie J, Scarr E (2002) Decreased muscarinic(1) receptors in the dorsolateral prefrontal cortex of subjects with schizophrenia. Mol Psychiatry 7: 1083–1091
5. Mahmood T, Yang P-C (2012) Western Blot: technique, theory, and trouble shooting. N Am J Med Sci 4:429–434
6. Koussounadis A, Langdon SP, Um IH, Harrison DJ, Smith VA (2015) Relationship between differentially expressed mRNA and mRNA-protein correlations in a xenograft model system. Sci Rep 5:10775
7. Pradidarcheep W, Stallen J, Labruyère WT, Dabhoiwala NF, Michel MC, Lamers WH (2009) Lack of specificity of commercially available antisera against muscarinergic and adrenergic receptors. Naunyn Schmiedeberg's Arch Pharmacol 379:397–402
8. Hamdani N, van der Velden J (2009) Lack of specificity of antibodies directed against human beta-adrenergic receptors. Naunyn Schmiedeberg's Arch Pharmacol 379:403–407
9. Ghosh R, Gilda JE, Gomes AV (2014) The necessity of and strategies for improving confidence in the accuracy of western blots. Expert Rev Proteomics 11:549–560
10. Hansen ES, Knudsen J (1985) Limitations in the use of the enzyme-linked immunosorbent assay (ELISA) for identification and quantification of thermogenin. J Immunol Methods 77: 297–304
11. Werry EL, Kassiou M (2022) Measuring affinity of ligands to the oxytocin receptor using radioligand binding. In: Werry EL, Reekie TA, Kassiou M (eds) Oxytocin: methods and protocols. Springer US, New York
12. Scarr E, Dean B (2008) Muscarinic receptors: do they have a role in the pathology and treatment of schizophrenia? J Neurochem 107: 1188–1195
13. Dean B, Crook JM, Opeskin K, Hill C, Keks N, Copolov DL (1996) The density of muscarinic M1 receptors is decreased in the caudate-putamen of subjects with schizophrenia. Mol Psychiatry 1:54–58
14. Dean B, Hayes W (1996) Decreased frontal cortical serotonin 2A receptors in schizophrenia. Schizophr Res 21:133–139
15. Sekiguchi H, Pavey G, Dean B (2019) Altered levels of dopamine transporter in the frontal pole and dorsal striatum in schizophrenia. NPJ Schizophr 5:20
16. Dean B, Pavey G, Thomas D, Scarr E (2006) Cortical serotonin (7, 1D) and (1F) receptors: effects of schizophrenia, suicide and antipsychotic drug treatment. Schizophr Res 88: 265–274
17. Jeon WJ, Gibbons AS, Dean B (2013) The use of a modified [3H]4-DAMP radioligand binding assay with increased selectivity for muscarinic M3 receptor shows that cortical CHRM3 levels are not altered in mood disorders. Prog Neuropsychopharmacol Biol Psychiatr 47:7–12
18. Dean B, Crossland N, Boer S, Scarr E (2008) Evidence for altered post-receptor modulation of the serotonin 2a receptor in schizophrenia. Schizophr Res 104:185–197
19. Dean B, Pavey G, Opeskin K (1997) [3H] raclopride binding to brain tissue from subjects with schizophrenia: methodological aspects. Neuropharmacology 36:779–786
20. Dean B, Soulby A, Evin GM, Scarr E (2008) Levels of [(3)H]pirenzepine binding in Brodmann's area 6 from subjects with schizophrenia is not associated with changes in the transcription factor SP1 or BACE1. Schizophr Res 106: 229–236

21. Crook JM, Tomaskovic-Crook E, Copolov DL, Dean B (2001) Low muscarinic receptor binding in prefrontal cortex from subjects with schizophrenia: a study of Brodmann's areas 8, 9, 10, and 46 and the effects of neuroleptic drug treatment. Am J Psychiatr 158:918–925
22. Rodbard D (1981) Mathematics and statistics of ligand assays: an illustrated guide. In: Langan J, Clapp JJ (eds) Ligand assay: analysis of international developments on isotopic and nonisotopic immunoassay. Masson Punlishing USA, Inc, New York
23. Pavey GM, Copolov DL, Dean B (2002) High-resolution phosphor imaging: validation for use with human brain tissue sections to determine the affinity and density of radioligand binding. J Neurosci Method 116:157–163
24. Dean B, Opeskin K, Pavey G, Naylor L, Hill C, Keks N et al (1995) [3H]paroxetine binding is altered in the hippocampus but not the frontal cortex or caudate nucleus from subjects with schizophrenia. J Neurochem 64:1197–1202
25. Conn PJ, Christopoulos A, Lindsley CW (2009) Allosteric modulators of GPCRs: a novel approach for the treatment of CNS disorders. Nat Rev Drug Disc 8:41–54
26. Ma L, Seager MA, Wittmann M, Jacobson M, Bickel D, Burno M et al (2009) Selective activation of the M1 muscarinic acetylcholine receptor achieved by allosteric potentiation. Proc Natl Acad Sci U S A 106:15950–15955
27. Shirey JK, Brady AE, Jones PJ, Davis AA, Bridges TM, Kennedy JP et al (2009) A selective allosteric potentiator of the M1 muscarinic acetylcholine receptor increases activity of medial prefrontal cortical neurons and restores impairments in reversal learning. J Neurosci 29:14271–14286
28. Dean B, Hopper S, Conn PJ, Scarr E (2016) Changes in BQCA allosteric modulation of [3H]NMS binding to human cortex within schizophrenia and by divalent cations. Neuropsychopharmacology 41:1620–1628
29. Hopper S, Pavey GM, Gogos A, Dean B (2019) Widespread changes in positive allosteric modulation of the muscarinic M1 receptor in some participants with schizophrenia. Int J Neuropsychopharmacol 22:640–650
30. Bradford MM (1976) A rapid and sensitive method for the quantitation of microgram quantities of protein utilizing the principle of protein-dye binding. Anal Biochem 72:248–254
31. Dean B, Pavey G, Chai SY, Mendelsohn FAO (1999) The localisation and quantification of molecular changes in the human brain using in situ radioligand binding and autoradiography. In: Dean B, Kleinman JE, Hyde TM (eds) Using CNS tissue in psychiatric research: a practical guide. Harwood Academic Press, Amsterdam
32. Murphree SS, Saffitz JE (1987) Quantitative autoradiographic delineation of the distribution of beta-adrenergic receptors in canine and feline left ventricular myocardium. Circ Res 60:568–579
33. Timp W, Timp G (2020) Beyond mass spectrometry, the next step in proteomics. Science. Advances 6:eaax8978

Chapter 2

Enhanced Bystander BRET (ebBRET) Biosensors as Biophysical Tools to Map the Signaling Profile of Neuropsychiatric Drugs Targeting GPCRs

Supriya A. Gaitonde and Michel Bouvier

Abstract

Bioluminescence resonance energy transfer (BRET) is a non-radiative energy transfer between a bioluminescent donor and a fluorescent acceptor with far-reaching applications in detecting physiologically relevant protein-protein interactions. The recently developed enhanced bystander BRET (ebBRET) biosensors have made it possible to rapidly determine the signaling profile of a series of ligands across a large number of GPCRs and their signaling repertoires, which has tremendous implications in the drug discovery process. Here we describe BRET and the ebBRET biosensors as investigational tools in establishing functional selectivity downstream of GPCRs.

Key words GPCR, Functional selectivity, Signaling profile, BRET, Enhanced bystander BRET (ebBRET), Biosensors, Antipsychotics

1 Introduction

G protein-coupled receptors (GPCRs) regulate diverse physiological processes and are the targets of nearly 30% of the currently marketed drugs, including but not limited to antipsychotic drugs, β-blockers, anti-asthmatics, and antihistamines [1–3]. They are capable of engaging multiple G proteins and β-arrestin, and the selectivity among these different effectors is highly influenced by the binding ligand stabilizing distinct conformations of the receptor [4–7]. This ligand-directed signaling bias or functional selectivity can be harnessed for the development of safer drugs by the selective activation of the therapeutically relevant signaling pathways while avoiding the pathways causing side effects [8–16].

On account of being evolutionarily conserved, the binding pockets of aminergic GPCRs bear striking similarities to one another causing promiscuous binding of ligands, also known as polypharmacology. Although this more often than not leads to

Leyre Urigüen and Rebeca Diez-Alarcia (eds.), *Schizophrenia: Methods and Protocols*, Methods in Molecular Biology, vol. 2687, https://doi.org/10.1007/978-1-0716-3307-6_2,

undesirable off-target effects [17–19], for multifactorial disorders such as neuropsychiatric conditions, there is much evidence in favor of polypharmacology [20, 21]. In particular many of the antipsychotic drugs bind to multiple monoaminergic receptors with varying degrees of affinity [21]. However, the indiscriminate targeting of multiple GPCRs can lead to devastating side effects; for example, in the context of schizophrenia, GPCRs such as the dopamine D_2, serotonin 5-HT_{2C}, muscarinic M_3, histamine H_1, and α_1 adrenergic receptors are notorious for causing antipsychotic treatment-associated side effects [22]. Just as not all subtypes of a GPCR subfamily lead to beneficial effects, not all signaling pathways downstream of a particular GPCR might be involved in therapeutic activity. The development of safer drugs thus necessitates segregating not only the receptors but also the pathways leading to the beneficial effects from those causing side effects. One of the techniques that can be employed to monitor the distinct pathways engaged by different drugs in living cells is bioluminescence resonance energy transfer (BRET).

BRET is a biophysical technique based on the non-radiative energy transfer between a bioluminescent protein isolated from marine organisms such as *Renilla reniformis* (BRET donor: Renilla luciferase or RlucII) and a fluorophore such as green fluorescent proteins (GFP) and their mutant variants (BRET acceptors) upon the oxidation of the bioluminescent enzyme substrate [23, 24]. BRET has been used to quantify protein-protein interactions and has been thoroughly used to explore the molecular pharmacology of GPCRs, including activation, engagement of different G proteins, recruitment of β-arrestin, desensitization, internalization, recycling, dimer and higher order complex formation, and for the demonstration of signaling through distinct intracellular compartments [24–32]. The dependence on distance (BRET is inversely proportional to the sixth power of the distance between the energy donor and acceptor) and the parallel orientation of the dipoles contributes to the physiological relevance of the observed protein-protein interactions. The fact that it is independent of external light stimulation eliminates chances of autofluorescence or direct stimulation of the acceptor as in FRET [33]. BRET can be conducted using intact mammalian cells in real time and in all cellular compartments, providing an edge over other biochemical techniques that require cell disruption to monitor protein-protein interactions such as co-immunoprecipitation or GST pull-down or measurement of second messenger levels.

Over the past two decades, different families of BRET-based biosensors have been developed by several groups including our own, whereby the BRET donor and acceptor have been fused to different pairs of potentially interacting protein partners. For example, the dissociation of the Gα and the Gβγ complex by detecting a decrease in BRET between Gα-RlucII and Gγ-GFP10 [27] initially serving as a proxy for GPCR activation or an increase in BRET upon recruitment of β-arrestin-RlucII to the GPCR tagged to EYFP ($BRET^1$) [34, 35] (Fig. 1) have been extensively used. As a step toward determining ligand-biased signaling and to rapidly screen a panel of wild-type Gα subunits activated by a given ligand-GPCR pair, the competition of the wild-type Gα subunit and GRK2 to interact with the Gβγ complex was harnessed to develop a biosensor that could detect an increase in BRET upon the recruitment of GRK2-GFP10 toward Gγ-RlucII; the lack of perturbation of the structure of Gα subunit is a major advantage, although the Gβγ complex (Gγ-RlucII) is modified (Fig. 1) [28, 36].

Recently, the natural higher affinity of the energy donor and acceptor from the same species *Renilla reniformis* (RlucII and rGFP) was tapped into for better efficiency of the resonance energy transfer [37], and a new suite of biosensors was developed, whereby the translocation of a protein tagged with RlucII to a cellular compartment to which rGFP was targeted (enhanced bystander BRET or ebBRET) can be determined [29, 38] (*see* **Note 1**). This was used to monitor G protein activation without modifying the heterotrimeric G protein complex by measuring the translocation of RlucII-fused G protein effectors to the plasma membrane. The responses translate into an increase in BRET upon the recruitment of Gα family-specific downstream effectors (tagged to RlucII) to the plasma membrane (rGFP is targeted to the plasma membrane); the biosensors are thus named "effector membrane translocation assay" (EMTA) biosensors [38] (*see* **Note 2**). While the recruitment of the donor-tagged effector reflects activation of a family of Gα subunits ($G\alpha_{i/o/z}$, $G\alpha_q$, or $G\alpha_{12/13}$), the specific member of the family activated can be unveiled by overexpressing individual wild-type Gα subunits. In addition to the plasma membrane, the acceptor has been successfully targeted to other cellular compartments such as the endosomes [29, 32], which has helped assess the dynamics of the GPCR activation in different compartments.

While BRET as a laboratory technique has been routinely adapted to the medium- to high-throughput platform in the past, the ebBRET-based biosensors have the distinct advantage of greater sensitivity and higher BRET signal efficiency contributed by the *Renilla reniformis*-derived BRET donor-acceptor pair (*see* **Note 3**). This makes it possible to rapidly determine an array of signaling pathways activated by a particular ligand and those activated by a library of compounds, thus aiding the process of mapping the signaling profile of therapeutically relevant molecules (*see* **Note 4**). In the following sections, the experimental details pertaining to the use of ebBRET EMTA biosensors to monitor the signaling profiles promoted by different drugs will be presented, using receptors that are targets for neuropsychiatric drugs such as the serotonin $5\text{-}HT_{2A}$ and muscarinic M_3 receptors.

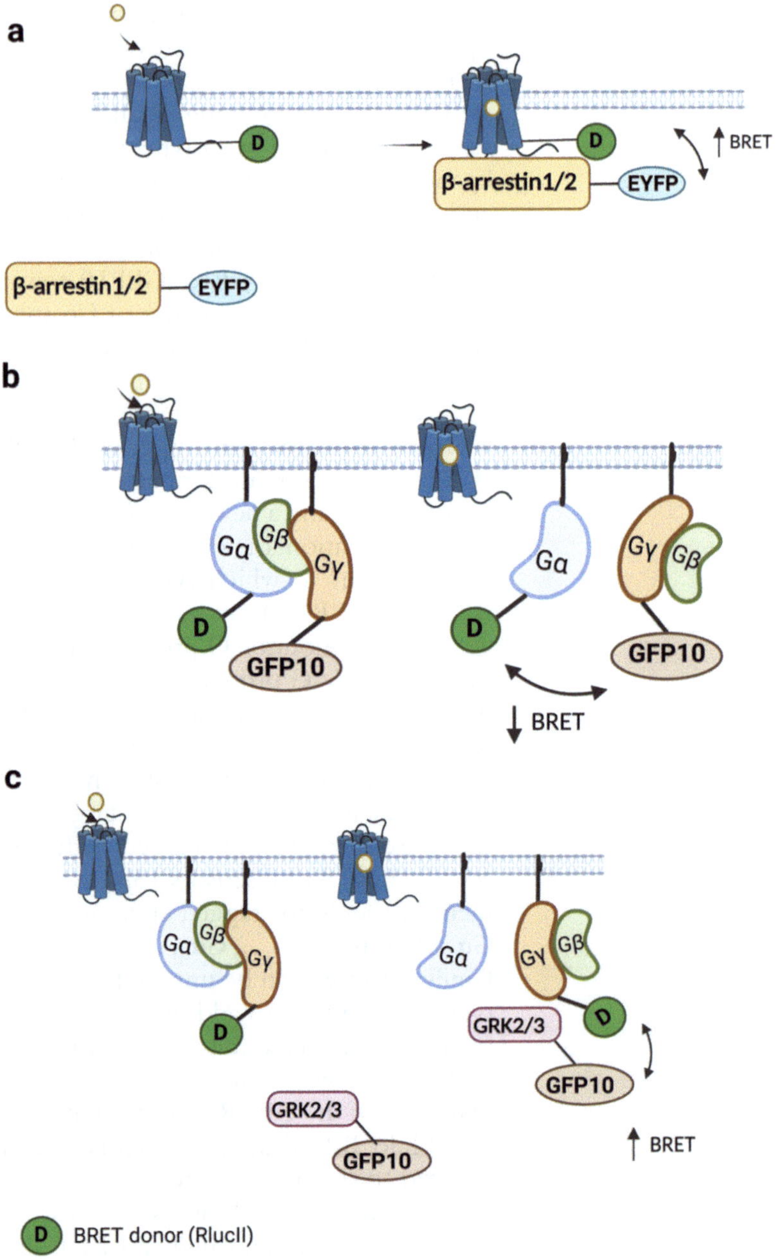

Fig. 1 Representation of the different BRET biosensors. (**a**) An increase in BRET upon recruitment of β-arrestin-EYFP to the GPCR tagged to RlucII is an indication of activation of the GPCR (BRET[1]). (**b**) A decrease in BRET upon the reorganization and/or dissociation of the heterotrimeric G protein complex (Gα-RlucII and Gγ-GFP10) indicates activation of the GPCR (BRET[2]). (**c**) Recruitment of GRK2-GFP10 to dissociated Gγ-RlucII leading to an increase in BRET signal is also used as a surrogate for activation (BRET[2]). (Created with BioRender.com)

2 Materials

2.1 Cell Culture

1. Cells: American-type cell culture (ATCC) human embryonic kidney 293 (HEK 293) cells.
2. Cell culture medium: Dulbecco's Modified Eagle Medium (DMEM) supplemented with 10% newborn calf serum (NCS) or fetal bovine serum (FBS) and 1% penicillin-streptomycin (100 U/mL penicillin and 100 μg/mL streptomycin).
3. Cell culture flasks and dishes: T150 cm^2 or T75 cm^2 flasks or 100 mm or 200 mm dishes.
4. Phosphate-buffered saline (PBS): 137 mM NaCl, 2.7 mM KCl, 8 mM Na_2HPO_4, and 2 mM KH_2PO_4.
5. Trypsin-EDTA solution: 0.025% trypsin and 0.01% EDTA in phosphate-buffered saline (PBS).
6. Transfection reagents: polyethyleneimine (PEI) (Alfa Aesar: catalog number: 43896) diluted in PBS (1 mg/mL).
7. 96-well white microplates: Greiner (catalog number: 655083) or Perkin Elmer (catalog number: 6005688).
8. Coating 96-well white microplates with poly-L-ornithine (Sigma Aldrich; catalog number: P4957): A stock solution of 1.5 mg/mL (100X) is prepared and stored at −20 °C and diluted (1X solution in PBS) when required such that for every well of a 96-well plate, 50 μL of the diluted solution (1X solution) is added. The poly-L-ornithine solution-treated cell culture plates are incubated at 37 °C for at least 24 h before washing them twice with PBS (75 μL/well). The PBS is aspirated and the plates are sealed in plastic bags and stored at 4 °C. Before use, the plates are brought back to room temperature and then incubated at 37 °C.
9. Plasmids and BRET biosensors: Plasmids encoding for the GPCRs, Gα subunits, and BRET donor-acceptor pair along with salmon sperm DNA (Invitrogen; catalog number: 15632011).

2.2 BRET Experiment

1. Tyrode's buffer: 137 mM NaCl, 0.9 mM KCl, 1 mM $MgCl_2$, 11.9 mM $NaHCO_3$, 3.6 mM NaH_2PO_4, 25 mM HEPES, 5.5 mM D-glucose, and 1 mM $CaCl_2$ (pH 7.4).
2. ebBRET substrates:

   DeepBlueC (coelenterazine 400a).

   Prolume Purple [methoxy e-Coelenterazine (Me-O-e-CTZ)].
3. BRET plate readers: Spark® multimode microplate reader or TriStar2 LB 942 Multidetection Microplate Reader or Mithras LB 940 microplate reader. Other instruments with the proper sensitivity and ability to monitor multiple wavelengths could also be used.

2.3 Data Analysis

Curve fitting and statistical analyses: GraphPad Prism software and Microsoft Excel.

3 Methods

3.1 Cell Culture

1. HEK293 cells are grown in DMEM supplemented with NCS or FBS (10%) and the antibiotics penicillin-streptomycin (1%). A 500 mL bottle of DMEM is supplemented with NCS or FBS (50 mL) and penicillin-streptomycin (5 mL). Cells can be grown in flasks (T150 cm^2 or T75 cm^2) or dishes (100 mm or 200 mm).
2. Cells are grown in incubators maintained at 37 °C with 5% CO_2 and 90% humidity.

3.2 Transfection

1. Prepare the DNA mix [comprising primarily of the plasmids encoding for the GPCR, Gα subunit, the G protein effectors tagged with the BRET donor (RlucII), and the plasma membrane anchor (the CAAX motif from KRas, which includes a prenylation and a polybasic domain) fused to the BRET acceptor rGFP (*see* **Note 5**)] in PBS (50 mL/1 μg of DNA).
2. Adjust the total quantity of the DNA to 1 μg with salmon sperm DNA (this total DNA amount is for a cell suspension of 1 mL and can be scaled up according to the experimental setup. For example, for a cell suspension of 2 mL, the volumes of all transfection components for 1 μg DNA plasmids, salmon sperm DNA, and PBS will be multiplied by 2).
3. Add the PEI solution to the DNA mix and incubate this mixture for at least 15 min. For 1 μg of DNA prepared in 50 μL of PBS, 50 μL of PEI solution is added. The dilute PEI solution is prepared in such a way as to maintain a DNA:PEI ratio of 1:3. For example, to prepare 50 μL of PEI, 3 μL of PEI (1 mg/mL) is added to 47 μL of PBS.
4. Wash the confluent cells (around 80% confluency) twice with PBS (10 mL), remove the PBS, and add 2 mL of trypsin to detach the cells. The cells treated with trypsin are incubated at 37 °C with 5% CO_2 and 90% humidity for 1 min.
5. Add DMEM (8 mL for a T75 cm^2 flask and 13 mL for a T150 cm^2 flask) and suspend the cells well.
6. Count the cells with a hematocytometer (take a 12 μL aliquot of the suspension from point 5) to determine the cell density.
7. Prepare a suspension of 3.5×10^5 cells/mL in DMEM. The total volume of this solution will depend on the specific requirements of the experiment. For example, to prepare a 50 mL cell suspension at 3.5×10^5 cells/mL from a suspension of cell

density 1×10^6 cells/mL, transfer 17.6 mL of cells from the 1×10^6 cells/mL cell suspension and add 32.5 mL DMEM to it.

8. Transfer the appropriate volume of the 3.5×10^5 cells/mL cell suspension to the DNA-PEI mixture. This volume will depend on the scale of the experiment. For example, for a DNA mixture of 1 μg, 1 mL of the cell suspension at 3.5×10^5 cells/mL is added to the DNA-PEI mix.
9. Immediately distribute 100 μL of this cell-DNA-PEI suspension in poly-L-ornithine-coated white 96-well microplates.
10. Incubate microplates for 48 h at 37 °C with 5% CO_2 and 90% humidity.

3.3 BRET Experiment

3.3.1 Treatment of the Cells

1. Aspirate the media of the 96-well plates on the day of the BRET experiment (48 h after transfection).
2. Using a multichannel pipette, add ~150 μL/well PBS solution to wash the cells. Cells are washed twice with PBS.
3. Aspirate the PBS, and then using a multichannel pipette, add Tyrode's buffer (80 μL/well).
4. Incubate cells at 37 °C in Tyrode's buffer for at least 15 min to equilibrate before proceeding with the BRET experiment.

3.3.2 BRET Measurement

1. Prepare serial dilutions of the drugs to be tested.
2. Distribute dilutions on the cells (10 μL) using multichannel pipettes.
3. Incubate the microplates with the drug-treated cells at 37 °C for at least 5 min.
4. Add the luciferase substrate (10 μL; 1.3 μM for Prolume Purple and 2.5 μM for DeepBlueC; *see* **Note 6**) and incubate for 5 min.
5. Measure the BRET signal on the microplate reader equipped with the following filters: 410/80 nm (donor; RlucII) and 515/40 nm (acceptor; rGFP).
6. Compounds can be screened in their agonist and antagonist mode in the same experimental plate to minimize variability. For the agonist mode, add the drug and measure BRET 10 min after and compare it with the BRET signal obtained with the vehicle treatment. For the antagonist mode, after the first read (agonist mode), treat the cells with the EC_{80} concentration of the reference agonist (10 μL) (the EC_{80} concentration of the reference agonist at the different signaling pathways being monitored is predetermined with concentration response curves at those pathways). The BRET signal obtained following a 5-min incubation with the reference agonist alone or in combination with the compound is compared.

3.3.3 Data Analysis

Raw BRET data is the ratio of the acceptor emission 515/40 nm over the emission of the donor 410/80 nm (rGFP/RlucII). Ligand-promoted BRET (ΔBRET) is determined by subtracting the BRET ratio of the vehicle condition from the BRET ratio of the ligand-treated conditions. To determine the potencies of the compounds ($logEC_{50}$ for agonists and $logIC_{50}$ for antagonists), concentration response curves can be constructed by treating the cells with increasing concentrations of the test compounds. To determine the relative efficacies (E_{max} values) of the compounds (partial agonist vs. full agonist), the ligand-promoted BRET (ΔBRET) obtained in the agonist mode can be normalized as a function of the maximal response of the reference agonist. Concentration response curves are obtained using a three-parameter logistic nonlinear regression model using GraphPad Prism.

Calculation of bias: The bias of a ligand across different signaling pathways (determined by the ebBRET biosensors) can be calculated by the method by Kenakin and Christopoulos [39, 40] from the transduction coefficient ($\log \tau/K_A$) or (Log R) generated from the concentration response curves on GraphPad Prism using the nonlinear regression parameter "New Operational Model with TauKa Ratios." The ($\log \tau/K_A$) value for a given pathway is first subtracted from that of a reference ligand (typically the endogenous ligand) to give the $\Delta(\log \tau/K_A)$, which is then subtracted from the $\Delta(\log \tau/K_A)$ obtained for a second signaling pathway to give the $\Delta\Delta(\log \tau/K_A)$ (such that $\Delta\Delta(\log \tau/K_A)$ is a positive value). The antilog of $\Delta\Delta(\log \tau/K_A)$ gives the bias factor. The $\Delta(\log \tau/K_A)$ can be used to compare the extent of activation of a series of ligands for a single signaling pathway.

4 Notes

1. ebBRET used for the EMTA sensors differs from the earlier BRET generation ($BRET^1$ and $BRET^2$; for details about the characteristics of $BRET^1$ and $BRET^2$, *see* Table 1) by the nature of the energy acceptor, which is rGFP rather than *Aequorea victoria* GFP. This takes advantage of the natural high affinity (micromolar range) between RlucII and rGFP and their co-evolution (both being from *Renilla reniformis*) yielding optimal energy transfer efficiency [37] and allows to gauge the translocation of either the GPCRs or their effectors to specific cellular compartments. Indeed, it has been shown that no spontaneous interactions occur between RlucII and rGFP when expressed in different compartments, but when co-localized in two-dimensional space within a compartment, they self-associate leading to a high transfer of energy with a bigger dynamic window than $BRET^1$ and $BRET^2$. This property is being used to monitor the translocation of cytoplasmic

Table 1
Wavelengths (nm) associated with the three BRET generations ($BRET^1$, $BRET^2$, and ebBRET)

SUBSTRATE (oxidized by RlucII)			FLUORESCENT ACCEPTORS							
$BRET^1$	$BRET^2$ and ebBRET		$BRET^1$		$BRET^2$				ebBRET	
Coel h[a]	DBC[b]	Prolume Purple	EYFP		GFP2		GFP10		rGFP	
Emission	Emission	Emission	Excitation	Emission	Excitation	Emission	Excitation	Emission	Excitation	Emission
480	400	405	511	530	400	510	400	510	480	508

[a]Coelenterazine h; [b]DeepBlueC (Coelenterazine 400a)

effectors to the G proteins at the plasma membrane. ebBRET uses the same bioluminescent substrates as $BRET^2$ (i.e., coelenterazine 400a also known as DeepBlueC or methoxy e-Coelenterazine [Me-O-e-CTZ] also known as Prolume Purple). A comparison of the emission wavelengths of the bioluminescence substrate/energy donor and the excitation and emission wavelengths of the fluorescent acceptors associated with $BRET^1$, $BRET^2$, and ebBRET is listed in Table 1.

2. The ebBRET biosensors have made it possible to monitor the direct activation of the different Gα subunits at the plasma membrane by monitoring the recruitment of Gα family-specific downstream effectors (Fig. 2 and Table 2) through BRET measurements between effector-RlucII and rGFP-CAAX. The identity of the activated Gα subunit can be revealed by overexpressing the individual unmodified Gα subunit of interest. Due to the lack of a soluble $G\alpha_s$-selective effector that is recruited from the cytosol to $G\alpha_s$ upon activation, we can take advantage of the fact that $G\alpha_s$ is released from the plasma membrane following activation [41, 42]. Hence to monitor $G\alpha_s$ leaving the membrane as a surrogate for its activation, RlucII is tagged to $G\alpha_s$ and a decrease in BRET between $G\alpha_s$-RlucII and rGFP-CAAX is monitored (Fig. 2). Taking advantage of the fact that more than 95% of GPCRs can activate either $G\alpha_{15}$ or $G\alpha_z$, a quasi-universal G15/Gz biosensor described in Avet, Mancini et al. (2020) allow to detect, in a single assay, if a given ligand activates any GPCR. Such a tool is particularly useful in high-throughput screening for orphan receptors in the context of early drug discovery process.
3. A higher efficiency of energy transfer [29, 37] and a greater dynamic window of activation with ebBRET as compared to the classical $BRET^2$ (GFP10) and $BRET^1$ (Venus) assays has been previously reported for detection of β-arrestin recruitment and GPCR endocytosis [29]. Figure 3 illustrates the greater dynamic window for the detection of $G\alpha_q$ activation by the $5\text{-}HT_{2A}$ receptor with the ebBRET GEMTA biosensor p63-RlucII/rGFPCAAX (Fig. 3a) compared to the biosensor detecting such activation using the competition between $G\alpha_q$ and GRK2-GFP10 for interacting with Gγ5-RlucII (Fig. 3b)

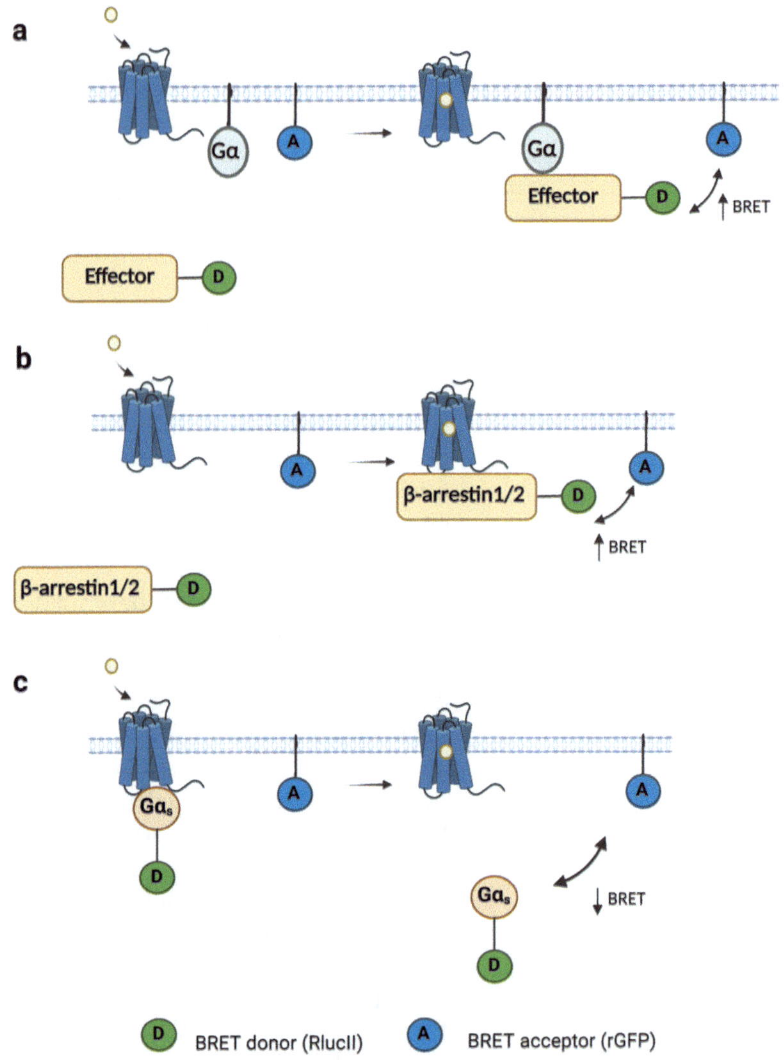

Fig. 2 Representation of the different ebBRET biosensors. (**a**) Recruitment to rGFP-CAAX attached to the plasma membrane of the Gα family-specific downstream effector fused with RlucII results in an increase in BRET signal indicating activation of the endogenous Gα family member or the overexpressed Gα subunit tested. (**b**) An increase in BRET due to greater proximity between β-arrestin-RlucII and rGFP-CAAX indicates ligand-mediated recruitment of β-arrestin to the receptor. (**c**) Activation of $G\alpha_s$ is associated to its leaving the plasma membrane leading to a decrease in BRET between $G\alpha_s$-RlucII and rGFP-CAAX. (Created with BioRender.com)

and the biosensor detecting the dissociation of the Gα (Gα-RlucII) from Gβγ (Gγ-GFP10) (Fig. 3c). The greater assay window obtained using the ebBRET GEMTA biosensor can be seen when reporting the absolute data or following normalization (%) as a function of the non-stimulated signal.

Table 2
The effectors associated with the GEMTA ebBRET biosensors for specific G protein subtypes

Gα subunits	Effector tagged to RlucII
$G\alpha_q$	p63-RhoGEF[a]
$G\alpha_{i/o/z}$	Rap1Gap[a]
$G\alpha_{12/13}$	PDZRhoGEF[a] and p115RhoGEF[b]

[a] [38]; [b] [44]

The robust signal is advantageous for high-throughput screening (*see* **Note 7**).

4. The ebBRET biosensors can be used to detect differences in the activation profile of a ligand and the potential bias of such ligands for subfamilies of Gα subunits by overexpressing the respective Gα subunits to be tested. The transduction coefficient ($\log \tau/K_A$) or (Log R) value for each pathway generated from the concentration response curves computed on GraphPad Prism can be used to calculate the bias factor compared to a reference compound as described in the "Methods" section. Figure 4 illustrates differences in the activation of the different $G\alpha_q$ family members through 5-HT_{2A} and M_3 receptors by their cognate ligands (5-HT and acetylcholine) detected with the GEMTA biosensor p63-RlucII/rGFP-CAAX, upon overexpressing the respective Gα subunits. The tables in Fig. 4c, d show the activation profile of the endogenous agonists 5-HT and acetylcholine for their respective receptors. The activation of a specific pathway by the two receptors was compared using the operational model as reported by Nagi and Pineyro [40] to get the ($\log \tau/K_A$) value for each pathway at the two receptors. As depicted in Fig. 4c, d, the rank order of efficacy for the different $G\alpha_q$ family members is clearly different for the 5-HT_{2A} and M_3 receptors. Indeed, whereas $G\alpha_q$ and $G\alpha_{14}$ are activated to the same extent by the M_3 receptor and represented the subtypes most efficiently activated by this receptor, $G\alpha_{14}$ and $G\alpha_q$ were activated to different extents by 5-HT_{2A}, $G\alpha_q$ being the subtype that is activated to the lowest extent by this receptor.

5. Assay development: As in the case of previously reported BRET-based biosensors [35], the optimum transfection conditions would be a culmination of titration experiments for a given BRET donor-acceptor pair. In such experiments, a fixed amount of the donor plasmid is expressed (minimum amount giving a reliably detectable bioluminescence signal) while expressing increasing amounts of the acceptor plasmid. For specific BRET interactions, a graph of the BRET signal against

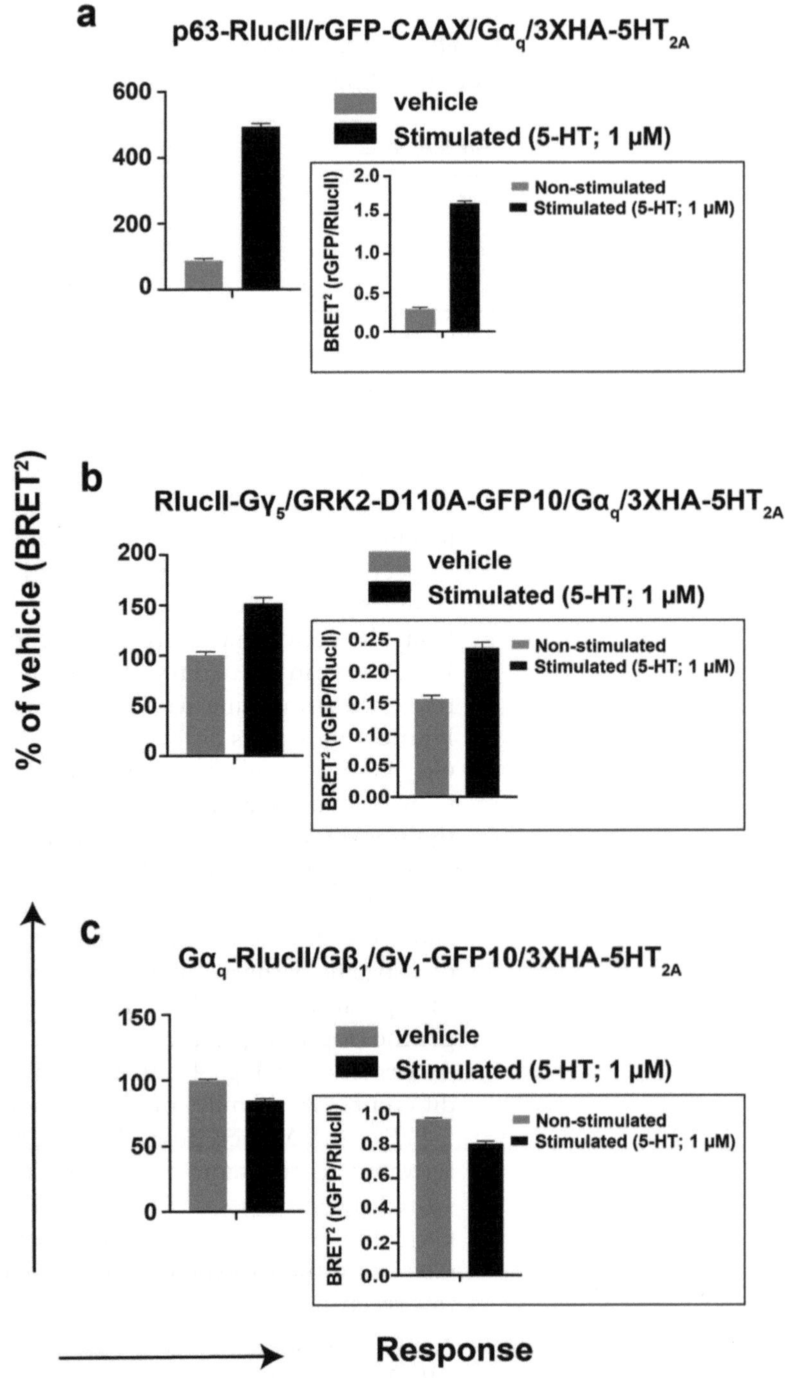

Fig. 3 Comparison of the dynamic windows obtained with the ebBRET (GEMTA), GRK2 competition and $G\alpha\beta\gamma$ dissociation biosensors (**a**) Activation of $G\alpha_q$ with 5-HT (1 μM; 15 min) in HEK 293 cells overexpressing the 3XHA-5-HT_{2A} receptor and the biosensor (p63-RlucII/rGFP-CAAX) and $G\alpha_q$. An increase in BRET signal indicates activation. (**b**) Activation of $G\alpha_q$ with 5-HT (1 μM; 15 min) in HEK 293 cells overexpressing the 3XHA-5-HT_{2A} receptor and the

the donor (bioluminescence)/acceptor (fluorescence) levels would depict a hyperbolic curve indicating that at a certain concentration (maximal) of the acceptor plasmid, all the donor molecules will be engaged [35]. In the case of the ebBRET EMTA biosensors, in addition to the optimization of the donor-acceptor transfection conditions, the coupling efficiency of a GPCR for a particular Gα subunit would also need to be considered, and titration experiments of the GPCR and the Gα subunit need to be performed. Keeping all components of the transfection mix constant, increasing amounts of either the GPCR or the Gα subunit plasmids are included. These experiments are particularly critical for non-canonical Gα subunits where possible competition for coupling by endogenously expressed canonical Gα subunits cannot be ruled out and overexpressing the most favorable amount of these subunits would help avoid overlooking the possible activation of such pathways. It should be emphasized that detection of a particular G protein coupling in heterologous systems reflects the possible couplings of a GPCR that needs to be validated in native systems to assess its physiological relevance.

6. Choice of the luciferase substrate: Similar to $BRET^2$, ebBRET utilizes DeepBlueC (coelenterazine 400a) and Prolume Purple [methoxy e-Coelenterazine (Me-O-e-CTZ)] as the luciferase substrates with RlucII, and the choice of the substrate selected depends on the assay to be conducted and the experimental conditions. Given that Prolume Purple has a ~ 6 times higher luminescence than DeepBlueC [43], it would be the substrate of choice for experiments requiring extended time-points.
7. Potential for medium- to high-throughput screening: ebBRET biosensors have recently been used to perform medium-throughput screening in 384-well plate format [29]. The results showed the robustness and reproducibility of the assay with a Z' value of 0.64–0.83 for AT_1R translocation to early endosome (endocytosis using AT_1R-RlucII and rGFP-FYVE). The landmark paper describing the ebBRET GEMTA

Fig. 3 (continued) biosensor (RlucII-$G_{\gamma5}$/GRK2-D110A-GFP10, $G\beta_1$ and $G\alpha_q$; the D110A mutant lacking the RGS domain having affinity for $G\alpha_q$ was used to eliminate direct interaction with $G\alpha_q$). An increase in BRET signal indicates activation. (**c**) Activation of $G\alpha_q$ with 5-HT (1 μM; 15 min) in HEK 293 cells overexpressing the 3XHA-5-HT_{2A} receptor and the biosensor ($G\alpha_q$-RlucII/$G\beta_1$/$G_{\gamma1}$-GFP10). A decrease in BRET signal indicates activation. Results are normalized with respect to the non-stimulated (vehicle) condition (mean ± SEM; $n = 3$) and the raw BRET data presented in inset. The 3XHA-tag appended at the N-terminus of the receptor allows for monitoring the receptor expression level at the cell surface by ELISA or FACS

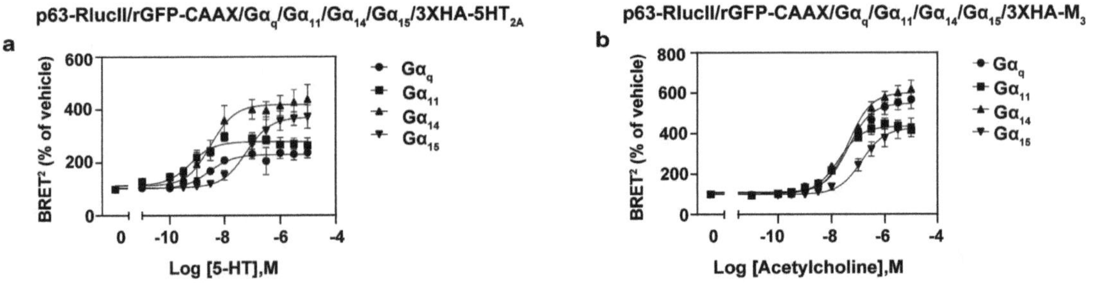

c

3XHA-5-HT$_{2A}$ (5-HT)		
Gα subunit	logEC$_{50}$ (± SEM)	Maximal response (± SEM) (% of vehicle)
Gα$_q$	-8.83 ± 0.18	200.40 ± 17.41
Gα$_{11}$	-9.07 ± 0.10	430.40 ± 20.86
Gα$_{14}$	-8.65 ± 0.08	608.80 ± 23.88
Gα$_{15}$	-7.48 ± 0.08	302.30 ± 7.83

d

3XHA-M$_3$ (Acetylcholine)		
Gα subunit	logEC$_{50}$ (± SEM)	Maximal response (± SEM) (% of vehicle)
Gα$_q$	-7.38 ± 0.10	437.10 ± 15.30
Gα$_{11}$	-7.74 ± 0.08	332.10 ± 9.21
Gα$_{14}$	-7.37± 0.08	493.00 ± 13.77
Gα$_{15}$	-6.87 ± 0.12	326.70 ± 18.48

e

	Log(τ/K_A)	
Gα subunit	3XHA-5-HT$_{2A}$ (5-HT)	3XHA-M$_3$ (Acetylcholine)
Gα$_q$	8.84 ± 0.20	7.10 ± 0.05
Gα$_{11}$	8.90 ± 0.35	6.89 ± 0.12
Gα$_{14}$	8.24 ± 0.16	7.34 ± 0.05
Gα$_{15}$	6.91 ± 0.34	6.45 ± 0.06

Fig. 4 Pathway-specific differences within a Gα family and signaling efficiency at G protein pathways. (**a**) Concentration response curves of 5-HT-promoted activation of the Gα$_q$ family members in HEK 293 cells overexpressing the 3XHA-5-HT$_{2A}$ receptor and the biosensor (p63-RlucII/rGFP-CAAX) and Gα$_q$, Gα$_{11}$, Gα$_{14}$ or Gα$_{15}$. (**b**) Concentration response curves of acetylcholine-promoted activation of the Gα$_q$ family members in HEK 293 cells overexpressing the 3XHA-M$_3$ receptor and the biosensor (p63-RlucII/rGFP-CAAX) and Gα$_q$, Gα$_{11}$, Gα$_{14}$ or Gα$_{15}$. Results are normalized with respect to the non-stimulated (vehicle) condition (mean ± SEM; $n = 3$). (**c**, **d**) Potencies (logEC$_{50}$) and efficacies (E$_{max}$) of 5-HT and acetylcholine to promote the activation of the Gα$_q$ family members through 3XHA-5-HT$_{2A}$ and 3XHA-M$_3$, respectively. (**e**) Comparison of the transduction coefficient (log τ/K_A) for activation of the Gα$_q$ family members by 5-HT (at 3XHA-5-HT$_{2A}$) and acetylcholine (at 3XHA-M$_3$)

biosensors reported the signaling profile of the prototypical agonists for 100 GPCRs [38] enabling a direct comparison of the coupling efficiency of the GPCRs for different Gα subunits and can serve as a resource for GPCR-guided drug discovery programs.

Conflict of Interest Statement

MB is the president of the scientific advisory board for Domain Therapeutics. MB has filed patent applications related to some of the biosensors described in this work, and the technology has been licensed to Domain Therapeutics. All biosensors are available for academic research through regular material transfer agreement. SAG declares no competing interests.

References

1. Nichols DE (2004) Hallucinogens. Pharmacol Ther 101:131–181
2. Roth BL (2019) Molecular pharmacology of metabotropic receptors targeted by neuropsychiatric drugs. Nat Struct Mol Biol 26:535–544
3. Velmurugan BK, Baskaran R, Huang C-Y (2019) Detailed insight on beta-adrenoceptors as therapeutic targets. Biomed Pharmacother 117:109039
4. Berg KA, Maayani S, Goldfarb J, Scaramellini C, Leff P, Clarke WP (1998) Effector pathway-dependent relative efficacy at serotonin type 2A and 2C receptors: evidence for agonist-directed trafficking of receptor stimulus. Mol Pharmacol 54:94–104
5. Urban JD, Clarke WP, von Zastrow M, Nichols DE, Kobilka B, Weinstein H et al (2007) Functional selectivity and classical concepts of quantitative pharmacology. J Pharmacol Exp Ther 320:1–13
6. Wenzel-Seifert K, Seifert R (2000) Molecular analysis of beta(2)-adrenoceptor coupling to G (s)-, G(i)-, and G(q)-proteins. Mol Pharmacol 58:954–966
7. Kenakin T (2019) Biased receptor signaling in drug discovery. Pharmacol Rev 71:267–315
8. Van Eps N, Altenbach C, Caro LN, Latorraca NR, Hollingsworth SA, Dror RO et al (2018) Gi- and Gs-coupled GPCRs show different modes of G-protein binding. Proc Natl Acad Sci U S A 115:2383–2388
9. Wacker D, Wang C, Katritch V, Han GW, Huang X-P, Vardy E et al (2013) Structural features for functional selectivity at serotonin receptors. Science 340:615–619
10. Masureel M, Zou Y, Picard L-P, van der Westhuizen E, Mahoney JP, Rodrigues JPGLM et al (2018) Structural insights into binding specificity, efficacy and bias of a beta2AR partial agonist. Nat Chem Biol 14: 1059–1066
11. Galandrin S, Oligny-Longpre G, Bouvier M (2007) The evasive nature of drug efficacy: implications for drug discovery. Trends Pharmacol Sci 28:423–430
12. Allen JA, Yost JM, Setola V, Chen X, Sassano MF, Chen M et al (2011) Discovery of beta-arrestin-biased dopamine D2 ligands for probing signal transduction pathways essential for antipsychotic efficacy. Proc Natl Acad Sci U S A 108:18488–18493
13. Monasky MM, Taglieri DM, Henze M, Warren CM, Utter MS, Soergel DG et al (2013) The beta-arrestin-biased ligand TRV120023 inhibits angiotensin II-induced cardiac hypertrophy while preserving enhanced myofilament response to calcium. Am J Physiol Heart Circ Physiol 305:H856–H866
14. Grim TW, Acevedo-Canabal A, Bohn LM (2020) Toward directing opioid receptor signaling to refine opioid therapeutics. Biol Psychiatry 87:15–21
15. Wingler LM, Lefkowitz RJ (2020) Conformational basis of G protein-coupled receptor signaling versatility. Trends Cell Biol 30:736–747
16. Kenakin T (2021) Biased signaling as allosteric probe dependence. Cell Signal 79:109844
17. Venkatakrishnan AJ, Deupi X, Lebon G, Tate CG, Schertler GF, Babu MM (2013) Molecular signatures of G-protein-coupled receptors. Nature 494:185–194
18. Gloriam DE, Foord SM, Blaney FE, Garland SL (2009) Definition of the G protein-coupled receptor transmembrane bundle binding pocket and calculation of receptor similarities for drug design. J Med Chem 52:4429–4442
19. Brianso F, Carrascosa MC, Oprea TI, Mestres J (2011) Cross-pharmacology analysis of G protein-coupled receptors. Curr Top Med Chem 11:1956–1963
20. Ramsay RR, Popovic-Nikolic MR, Nikolic K, Uliassi E, Bolognesi ML (2018) A perspective on multi-target drug discovery and design for complex diseases. Clin Transl Med 7:3
21. Morphy R, Kay C, Rankovic Z (2004) From magic bullets to designed multiple ligands. Drug Discov Today 9:641–651
22. Montastruc F, Palmaro A, Bagheri H, Schmitt L, Montastruc J-L, Lapeyre-Mestre M (2015) Role of serotonin 5-HT2C and histamine H1 receptors in antipsychotic-induced diabetes: a pharmacoepidemiological-pharmacodynamic study in VigiBase. Eur Neuropsychopharmacol 25:1556–1565
23. Xu Y, Piston DW, Johnson CH (1999) A bioluminescence resonance energy transfer (BRET) system: application to interacting circadian clock proteins. Proc Natl Acad Sci U S A 96:151–156
24. Angers S, Salahpour A, Joly E, Hilairet S, Chelsky D, Dennis M et al (2000) Detection of beta 2-adrenergic receptor dimerization in living cells using bioluminescence resonance energy transfer (BRET). Proc Natl Acad Sci U S A 97:3684–3689
25. Perroy J, Pontier S, Charest PG, Aubry M, Bouvier M (2004) Real-time monitoring of ubiquitination in living cells by BRET. Nat Methods 1:203–208

26. Galés C, Rebois RV, Hogue M, Trieu P, Breit A, Hébert TE et al (2005) Real-time monitoring of receptor and G-protein interactions in living cells. Nat Methods 2:177–184
27. Galés C, Van Durm JJJ, Schaak S, Pontier S, Percherancier Y, Audet M et al (2006) Probing the activation-promoted structural rearrangements in preassembled receptor-G protein complexes. Nat Struct Mol Biol 13:778–786
28. Masuho I, Ostrovskaya O, Kramer GM, Jones CD, Xie K, Martemyanov KA (2015) Distinct profiles of functional discrimination among G proteins determine the actions of G protein-coupled receptors. Sci Signal 8:ra123
29. Namkung Y, Le Gouill C, Lukashova V, Kobayashi H, Hogue M, Khoury E et al (2016) Monitoring G protein-coupled receptor and β-arrestin trafficking in live cells using enhanced bystander BRET. Nat Commun 7: 12178
30. Maziarz M, Park J-C, Leyme A, Marivin A, Garcia-Lopez A, Patel PP et al (2020) Revealing the activity of trimeric G-proteins in live cells with a versatile biosensor design. Cell 182:770–785.e16
31. Wright SC, Bouvier M (2021) Illuminating the complexity of GPCR pathway selectivity – advances in biosensor development. Curr Opin Struct Biol 69:142–149
32. Wright SC, Lukasheva V, Le Gouill C, Kobayashi H, Breton B, Mailhot-Larouche S et al (2021) BRET-based effector membrane translocation assay monitors GPCR-promoted and endocytosis-mediated G(q) activation at early endosomes. Proc Natl Acad Sci U S A 118
33. Milligan G (2004) Applications of bioluminescence- and fluorescence resonance energy transfer to drug discovery at G protein-coupled receptors. Eur J Pharm Sci 21:397–405
34. Bertrand L, Parent S, Caron M, Legault M, Joly E, Angers S et al (2002) The BRET2/arrestin assay in stable recombinant cells: a platform to screen for compounds that interact with G protein-coupled receptors (GPCRS). J Recept Signal Transduct Res 22:533–541
35. Hamdan FF, Percherancier Y, Breton B, Bouvier M. Monitoring protein-protein interactions in living cells by bioluminescence resonance energy transfer (BRET). Curr Protoc Neurosci. 2006.;Chapter 5:Unit 5.23
36. Mende F, Hundahl C, Plouffe B, Skov LJ, Sivertsen B, Madsen AN et al (2018) Translating biased signaling in the ghrelin receptor system into differential in vivo functions. Proc Natl Acad Sci U S A 115:E10255–E10264
37. Molinari P, Casella I, Costa T (2008) Functional complementation of high-efficiency resonance energy transfer: a new tool for the study of protein binding interactions in living cells. Biochem J 409:251–261
38. Avet C, Mancini A, Breton B, Le Gouill C, Hauser AS, Normand C et al (2022) Effector membrane translocation biosensors reveal G protein and βarrestin coupling profiles of 100 therapeutically relevant GPCRs. elife 11: e74101
39. Kenakin T, Christopoulos A (2013) Signalling bias in new drug discovery: detection, quantification and therapeutic impact. Nat Rev Drug Discov 12:205–216
40. Nagi K, Pineyro G (2016) Practical guide for calculating and representing biased signaling by GPCR ligands: a stepwise approach. Methods 92:78–86
41. Allen JA, Yu JZ, Donati RJ, Rasenick MM (2005) Beta-adrenergic receptor stimulation promotes G alpha s internalization through lipid rafts: a study in living cells. Mol Pharmacol 67:1493–1504
42. Martin BR, Lambert NA (2016) Activated G protein Gαs samples multiple endomembrane compartments. J Biol Chem 291:20295–20302
43. Kobayashi H, Picard L-P, Schönegge A-M, Bouvier M (2019) Bioluminescence resonance energy transfer-based imaging of protein-protein interactions in living cells. Nat Protoc 14:1084–1107
44. Avet C, Sturino C, Grastilleur S, Gouill CL, Semache M, Gross F et al (2020) The PAR2 inhibitor I-287 selectively targets Gα(q) and Gα(12/13) signaling and has anti-inflammatory effects. Commun Biol 3:719

Chapter 3

[^{35}S]GTPγS (Guanosine-5′-O-(γ-thio)triphosphate-[^{35}S]) Binding Scintillation Proximity Assay Experiments in Postmortem Brain Tissue

I. Muneta-Arrate and R. Diez-Alarcia

Abstract

Heterotrimeric guanine nucleotide-binding proteins (G proteins) are the very first effector in signal transduction events triggered by G-protein-coupled receptors (GPCRs). One of the most widely used approaches for determining GPCR activity in native tissue is based on the binding of [^{35}S]GTPγS. Classically, an heterogeneous procedure including a filtration step has been used, but a modification of the protocol including an immunoprecipitation step has allowed the specific discrimination of the contribution of the different Gα subunit subtypes to the effect of each ligand. Nowadays, that the concept of functional selectivity has been demonstrated for several ligands and GPCRs, information obtained from this methodological approach will be very useful for broadening the knowledge of GPCRs signaling profiles and describing the effect of different ligands over them. In this chapter we will describe the detailed protocol of antibody-capture [^{35}S]GTPγS scintillation proximity assay (SPA) in order to provide the reader with comprehensive guidelines to study receptor-mediated functional activation of different Gα-protein subtypes in native mammalian brain membranes. In addition, advantages and limitations of this method will be described, as well as future direction in the application of this approach indicated.

Key words G protein, [^{35}S]GTPγS binding, Immunoprecipitation, Brain, GPCR, SPA

1 Introduction

GPCRs (G-protein-coupled receptors), also known as seven transmembrane domain (7TM) receptors, form one of the largest membrane receptor subfamilies of plasma membrane proteins. These receptors bind to diverse type of ligands, such as ions, small molecules, and peptides, linking them to downstream signaling [1]. The 7TM receptor superfamily is also termed as GPCRs because it exerts effects in response to different ligands through the association with G proteins. G proteins are heterotrimeric guanine nucleotide-binding proteins, constituted by three subunits: α, β, and γ. G proteins mediate the earliest step in cell response to

Leyre Urigüen and Rebeca Diez-Alarcia (eds.), *Schizophrenia: Methods and Protocols*, Methods in Molecular Biology, vol. 2687, https://doi.org/10.1007/978-1-0716-3307-6_3,

external events, by linking cell surface receptors to intracellular signaling [2], and serve as transducers or amplifiers of signals from GPCRs to intracellular effectors. Moreover, GPCRs can also bind to other cytosolic adaptors, including β-arrestins [3].

Heterotrimeric G proteins have a crucial role in defining the specificity and temporal characteristics of cellular responses. Upon ligand activation, there is a conformational change of GPCRs that increases their affinity for G proteins, leading to G-protein recruitment. In the inactive conformation, Gα subunit is bound to GDP and to Gβγ heterodimer forming a heterotrimeric complex. After receptor activation, Gα subunit binds to GTP and dissociates from the Gβγ dimer [4]. The signal terminates when the GTP is hydrolyzed by the GTPase activity of Gα subunits, to GDP and phosphate, and the Gβγ complex binds to Gα, forming the inactive G protein [5, 6].

A direct evaluation of GPCR activation can be made by measuring the stimulation or inhibition of guanine nucleotide exchange on G-protein receptors using radiolabeled GTP analogues. This approach is called sulfur 35-labeled guanosine-5′-O-(γ-thio)-triphosphate ([^{35}S]GTPγS) binding assay. Measurement of G-protein activation is the functional consequence of receptor occupancy as earliest event, and it is not subjected to signal amplification [7, 8]. Conventional [^{35}S]GTPγS binding assays are limited to the measurement of GPCR coupling to $G\alpha_{i/o}$ proteins, probably due to their higher rates of nucleotide exchange and constitutive activity [9]. Nevertheless, different assays have been developed in order to measure different Gα-protein responses such as [^{35}S]GTPγS binding immunoprecipitation assay using magnetic beads or scintillation proximity assays (SPA) [10]. These assays combine classical [^{35}S] GTPγS binding assay with immunoprecipitation and can be applied for studying receptor of interest in cell cultures and in native tissues.

In this chapter we will describe the detailed protocol of antibody-capture [^{35}S]GTPγS scintillation proximity assay (SPA) in order to provide the reader with comprehensive guidelines to study receptor-mediated functional activation of different G-α-protein subtypes in native mammalian brain membranes. In addition, the advantages and limitations of this method will be described, and future direction in the application of this approach indicated.

2 Materials

Prepare all solutions using ultrapure water (prepared by purifying deionized water, to attain a sensitivity of 18 MΩ-cm at 25 °C) and analytical grade reagents. Prepare and store all reagents at 4–8 °C (unless indicated otherwise). Diligently follow all waste disposal regulations when disposing waste materials.

2.1 Samples

Mammalian brain samples (stored at −80 °C).

2.2 Equipment

Motor-driven Teflon/glass homogenizer (IKA Labortechnik or similar).

Sorvall RC-5C centrifuge with SM-24 rotor (or similar). Centrifugations at 1,000 and 40,000 × g and 4 °C are needed.

Microplate reader (ELx808 Biotek or similar).

Thermostatizated microplate shaker (ELMI Sky Line, DTS-4 or similar). Incubations at 350 and 400 rpm and at 22 and 30 °C are needed.

Refrigerated Eppendorf Centrifuge 5810R (microtube rotor F45–30-11/microplate rotor A-4-62) (or similar). Centrifugations at 1,000 and 21,000 × g and at 4 and 22 °C are needed.

MicroBeta microplate scintillation counter (Perkin Elmer) (or similar).

2.3 Solutions

Homogenization buffer: 50 mM Tris-HCl (pH 7.4), 3 mM $MgCl_2$, 1 mM ethylene glycol-bis(2-aminoethylether)-N,N,N′,N′-tetraacetic acid (EGTA) and 1 mM DL-dithiothreitol (DTT), 10 M sucrose. Store 4–8 °C. For 1 L of homogenization buffer, 6.06 g of Tris (MW 121.1), 0.61 g of $MgCl_2$ (MW 95.2 + 6 × 18 H_2O), 0.38 g of EGTA (MW 380.4), 0.15 g of DTT (MW 154.2), and 8.56 g sucrose (MW 324.3) must be dissolved in mQ H_2O. Adjust pH to 7.4 with HCl 1 N.

Centrifugation buffer: 50 mM Tris-HCl (pH 7.4), 3 mM $MgCl_2$, 1 mM EGTA, and 1 mM DTT. Store 4–8 °C. For 1 L of centrifugation buffer, 6.06 g of Tris (MW 121.1), 0.61 g of $MgCl_2$ (MW 95.2 + 6 × 18 H_2O), 0.38 g of EGTA (MW 380.4), and 0.15 g of DTT (MW 154.2) must be dissolved in mQ H_2O. Adjust pH to 7.4 with HCl 1 N.

Incubation buffer: 50 mM Tris-HCl (pH 7.4), 3 mM $MgCl_2$, 1 mM EGTA, 1 mM DTT, and 100 mM NaCl. Store 4–8 °C. For 1 L of incubation buffer, 6.06 g of Tris (MW 121.1), 0.61 g of $MgCl_2$ (MW 95.2 + 6 × 18 H_2O), 0.38 g of EGTA (MW 380.4), and 5.84 g of NaCl (MW 58.4) must be dissolved in mQ H_2O. Adjust pH to 7.4 with HCl 1 N. 1 mM DTT will be provided by the radioligand solution.

2 10^{-2} M GDP solution (stocked in −20 °C freezer in aliquots). 8.9 mg of guanosine diphosphate (GDP) (Sigma-Aldrich, G-7127) are dissolved for each mL of incubation buffer in order to get a *2* 10^{-2} M solution.

2 10^{-3} M GTPγS solution 1 mM GTPγS solution (stocked in-20 °C freezer in aliquots). 1.1 mg of guanosine 5’-O-[gamma-thio] triphosphate (GTPγS) (Sigma-Aldrich, G-8634) are dissolved for each mL of incubation buffer in order to get a 2 10^{-3} M solution.

Guanosine-5′-O-(γ-thio)-triphosphate ([^{35}S]GTPγS) (Perkin Elmer, NEG030-H, 1250 Ci/mmol): prepared to a concentration of 16 nM in 10 mM Tricine (pH 7.6) with 10 mM DTT solution and stocked in aliquots at −80 °C. For 20 ml of Tricine-DTT solution, 35.9 mg of Tricina OH (MW 179.2) and 31 mg de DTT (MW 154.2) must be dissolved in mQ H_2O. Adjust pH to 7.7 with NaOH 0.06 M.

Detergent solution: 1% v/v Igepal 1% + 0.1% w/v sodium docecyl sulfate (SDS) + 1% Protease Inhibitor Cocktail in incubation buffer (added freshly). For 100 mL of detergent solution, 1 mL of Igepal and 0.1 g of SDS (MW 288.4) must be dissolved in mQ H_2O. Store 4–8 °C.

15% BSA in incubation buffer (15 g of BSA in 100 ml of incubation buffer). Store 4–8 °C. Bovine serum albumin (BSA). Sigma-Aldrich.

2.4 Other Reagents

Antibodies (*see* Table 1)

Protein A-PVT SPA beads (Perkin Elmer, RPNQ0019)

2.5 Consumables

Centrifuge tubes

Microtubes (1.5 mL)

Bradford Bio-Rad Protein Assay (or similar)

Isoplate (Perkin Elmer Life Sciences, Maanstraat, Germany) (or similar). A 96-well plate with clear bottom allowing a final incubation volume higher than 250 μL and compatible with bottom luminescence detection

Seal Plate (Perkin Elmer Life Sciences or similar)

Table 1
Monoclonal and polyclonal antibodies used in [^{35}S]GTPγS binding scintillation proximity assay (SPA). GDP concentration and final concentration of antibodies used in these assays are also described

Target	Description	Ref.	Distributor	[GDP] (μM)	Final concentration
$G_{\alpha i1}$	Mouse monoclonal anti-$G_{\alpha i1}$	sc-56536	Santa Cruz	100 μM	0.1 μg/well
$G_{\alpha i2}$	Mouse monoclonal anti-$G_{\alpha i2}$	sc-13534	Santa Cruz	50 μM	0.1 μg/well
$G_{\alpha i3}$	Mouse monoclonal anti-$G_{\alpha i3}$	ABIN6258933	Antibodies on-line	100 μM	0.05 μg/well
$G_{\alpha o}$	Mouse monoclonal anti-$G_{\alpha o}$	sc-393874	Santa Cruz	50 μM	0.05 μg/well
$G_{\alpha q/11}$	Mouse monoclonal anti-$G\alpha_{q/11}$	sc-515689	Santa Cruz	10–50 μM	0.1 μg/well
$G_{\alpha S}$	Mouse monoclonal anti-$G_{\alpha s}$	sc-377435	Santa Cruz	100 μM	0.1 μg/well
$G_{\alpha z}$	Rabbit polyclonal anti-$G_{\alpha z}$	ABIN653561	Antibodies on-line	100 μM	0.12 μg/well
$G_{\alpha 13}$	Rabbit polyclonal anti-$G\alpha_{13}$	ABIN1537979	Antibodies on-line	100 μM	0.08 μg/well

2.6 Software

Software for curve-fitting program, GraphPad Prism

3 Methods

3.1 Membrane Preparation

1. Thaw tissue samples at 4 °C and homogenize with a glass/Teflon grinder (ten strokes at maximum speed) in 30 v/w of ice-cold homogenization buffer.
2. Centrifuge homogenates at 1,000 × g for 10 min at 4 °C (3,000 rpm with Sorvall RC-5C centrifuge, SM-24 rotor).
3. Discard pellets (P1 fraction) and carefully recover supernatants in clean tubes.
4. Re-centrifuge supernatants at 40,000 × g for 10 min at 4 °C (18,000 rpm with Sorvall RC-5C centrifuge, SM-24 rotor).
5. Discard supernatants and resuspend resultant pellets in 20 v/w of fresh cold centrifugation buffer with a glass stick. Re-centrifuge at 40,000 × g for 10 min at 4 °C (18,000 rpm with SorvallRC-5C centrifuge, SM-24 rotor).
6. Resuspend obtained pellets in five v/w of ice-cold centrifugation buffer.
7. Determine protein content by the method of Bradford with BSA as standard.
8. Aliquot membrane homogenate in 0.5–2 mg aliquots and finally centrifuge at 21,000 × g (14,000 rpm with Eppendorf 5810R centrifuge and F45-30-11 rotor) during 15 min at 4 °C (*see* **Note 1**).
9. Discard supernatant layer and store pellets at −80 °C until assay (*see* **Note 2**) (Fig. 1).

3.2 [^{35}S]GTPγS Binding, Immunoprecipitation, and Detection

Antibody-capture [^{35}S]GTPγS scintillation proximity assay (SPA) is based on the immunocapture of the specific Gα subunit subtype to be studied and the use of scintillant beads for the conversion of the radioligand particle emission into detectable photons [11, 12]. Different scintillant materials and coatings are available; thus different types of SPA beads can be tested in order to optimize each individual experimental setting. In the present protocol, the immunocapture of Gα subunits is mediated by the binding of Fc region of the IgG molecules of the antibodies with the protein A coating [13, 14] of polyvinyl toluene (PVT) beads (*see* **Notes 3** and **4**). This binding puts the [^{35}S]GTPγS bound to Gα in close proximity to the scintillating agent, PVT, that allows the conversion of the radioactive emission of [^{35}S] into a photon emission detectable by means of a scintillation counter (Fig. 2). This photon emission from the scintillant will only occur in cases where the radioligand is close enough to the beads; this is the [^{35}S]GTPγS bound to the α subunits that have been recognized by the specific

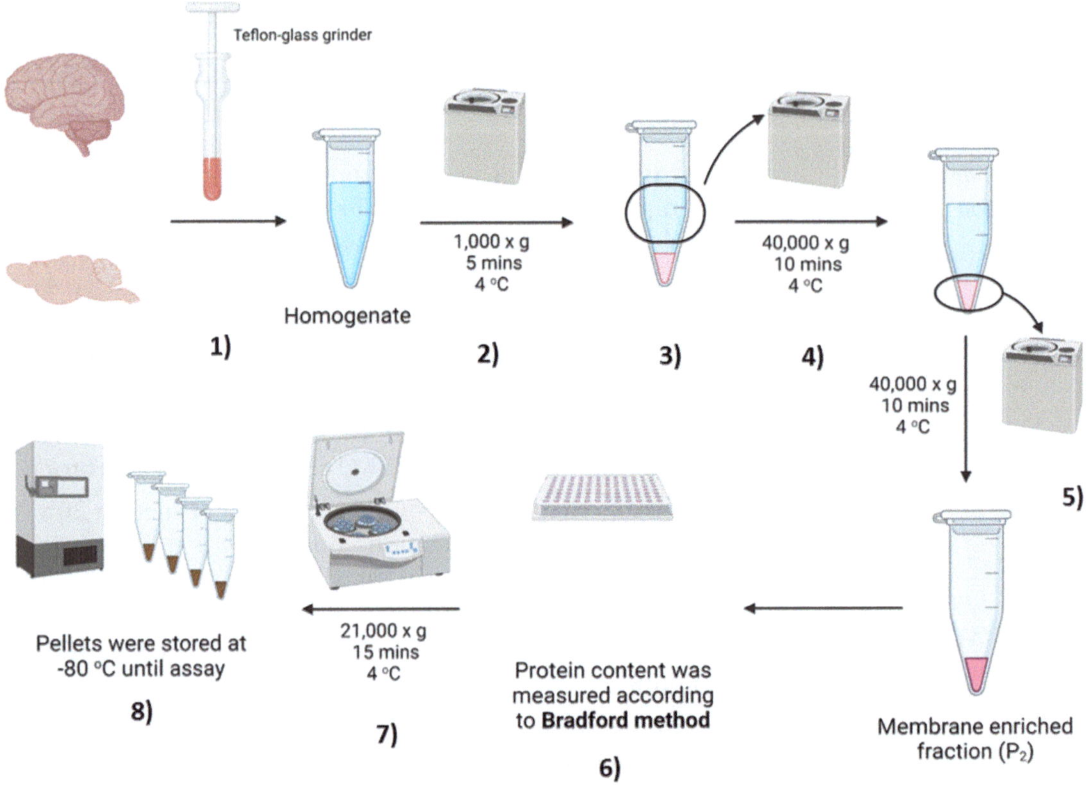

Fig. 1 Schematic representation of membrane-enriched fraction (P_2 fraction) obtaining process

antibody used [15] (*see* **Note 5**). The rest of [^{35}S]GTPγS, both free and bound to other Gα subunits, does not precipitate together with the beads and will not be detected [16].

In order to specifically measure the activation of a given type of Gα subunit triggered by a ligand, at least three different incubation conditions must be assayed: nonspecific binding (NSB), basal binding (BB), and binding in the presence of the ligand (L). Moreover, to determine the specific GPCR responsible for the observed effects, pharmacological antagonism should be used, carrying out co-incubation in the presence of selective antagonists of the receptors which role is suspected (ligand + antagonist incubation condition; L + A) (*see* **Note 6**).

Nonspecific binding (NSB) corresponds to the binding of [^{35}S]GTPγS in the presence of an excess of non-labeled ligand, 100 μM GTPγS. In this condition, GTPγS displaces the radioligand from the nucleotide-binding sites of the Gα subunits. Binding detected in this condition is supposed to be due to interaction with other elements present in the well (membranes, other proteins, etc.), binding to the Gα subunits in a different place than the nucleotide-binding pocket, or even residual detection of free [^{35}S]GTPγS. This value must be subtracted from that obtained for the rest of the conditions to determine the specific radioligand

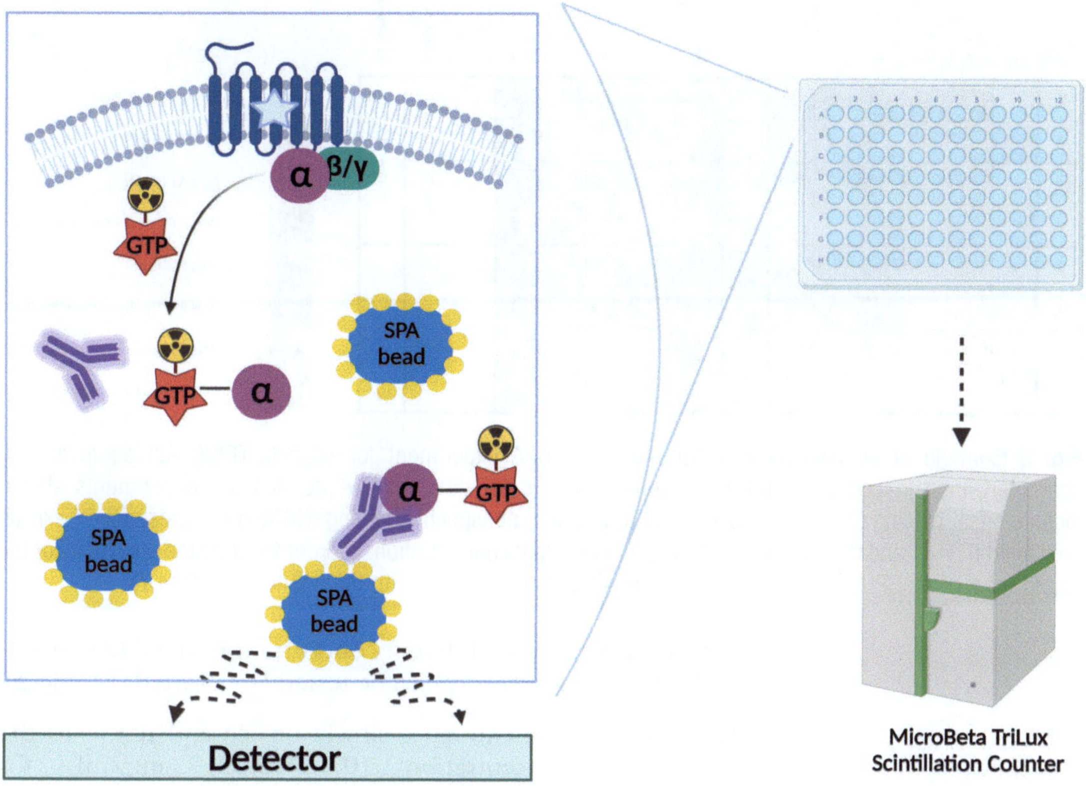

Fig. 2 Schematic representation of antibody-capture [^{35}S]GTPγS scintillation proximity assay (SPA). This figure depicts a single well of 96-well plate containing a membrane preparation bearing as example of an agonist-activated GPCR. The activated Gα subunit, released from the membrane after initial incubation with detergent, binds [^{35}S]GTPγS and is recognized by the primary anti-Gα antibody shown in gray. The measurement or radioactive energy is only limited to radioactive molecules that are held in close proximity to a polymer microsphere containing scintillant (SPA bead). The beads capture antibody-Gα protein complex. This interaction allows conversion of radioactivity to detectable photon emission by the scintillant and that is read by a microplate scintillation counter

binding values. Basal binding (BB) corresponds with the binding of [^{35}S]GTPγS to Gα subunits in the absence of any ligand, and it has been described to be different for each of them.

The final concentration of tested drug (L) and antagonist (A) to be used depends on the kind of assay. For concentration-effect experiments, a range of increasing concentration from 10^{-12} M to 10^{-3} M tends to be used. For characterization experiments or multiple subunit screening, one-point concentration at a submaximal concentration of the drug is used.

All these incubation conditions are tested in the same plate with several replicates per condition (Fig. 3).

Assays are performed in 96-well Isoplates (Perkin Elmer Life Sciences, Maanstraat, Germany) and in a final volume of 200 μL of incubation buffer plus 0.4 nM [^{35}S]GTPγS, 11–15 μg of protein

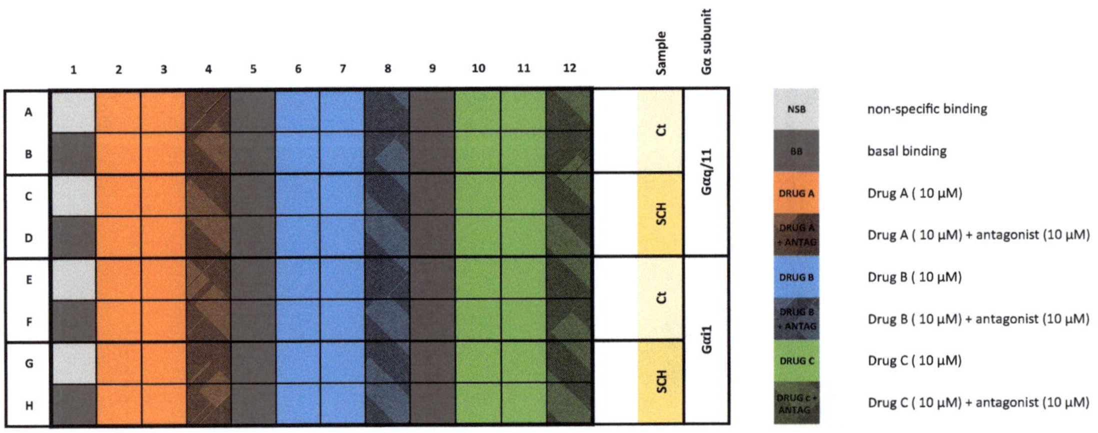

Fig. 3 Example of 96-well plate distribution of a SPA experiment for specific GPCR functional activity characterization in control and schizophrenia subjects for different Gα proteins. All experiments should include basal binding (BB) in the absence of any ligand, nonspecific binding (NBS) in the presence of excess unlabeled GTPγS, and the ligand (L) of interest at different concentration. In order to study the specificity of the ligand for a given GPCR, selective antagonist will be co-incubated with the ligand (*see* **Note 7**)

per well (*see* **Note 7**), and different concentrations of GDP depending on the Gα subunit subtype to be tested (*see* **Note 8**) (Table 2).

1. Thaw the brain membranes slowly on ice and dilute to the appropriate concentration (0.067–0.091 mg/mL for 11–15 μg protein/well) with ice-cold incubation buffer (*see* **Note 7**).
2. Prepare the reagent solutions according to the design of the experiment and the different incubation conditions to be carried out: GDP (dependent on the Gα subunit subtype to be studied) (*see* **Note 8**), drug to be tested (one-point concentration vs complete concentration-response experiments) (*see* **Note 9**), antagonist, vehicle, etc. (*see* **Note 10**). Stock 2 10^{-3} M GTPγS aliquots are directly used.
3. In case of experiments including co-incubation of tested drug in the presence of an antagonist, brain membranes are preincubated for 15 min at 30 °C and gentle orbital shaking in the presence of GDP solution, GTPγS solution (in case of NSB), antagonist, and/or vehicle (for BB and L conditions).
4. Then, drugs of interest are added to the corresponding wells and a second preincubation for 30 min at 30 °C and gentle orbital shaking is carried out.
5. Finally, 5 μL of the radioligand [^{35}S]GTPγS 16 nM are added, final concentration of 0.4 nM, and incubate for 90 min at 30 °C and gentle orbital shaking.
6. After this incubation, 20 μL of detergent solution (1% Igepal, 0.1% sodium deoxycholate, 1% protease inhibitor cocktail, freshly added) are added to each well to solubilize membranes

Table 2
General protocol of [^{35}S]GTPγS binding combined with scintillation proximity assay (SPA). BB, basal binding (binding values in the absence of any drug); NBS, nonspecific binding (defined as the remaining [^{35}S]GTPγS binding in the presence of 100 μM unlabeled GTPγS) (*see* Notes 11–13)

	BB	L	L+ A	NSB	[]final
GDP	10 μL	10 μL	10 μL	10 μL	1/10/50/100/200 μM
Vehicle	20 μL	10 μL	–	10 μL	
Incubation buffer	–	–	–	–	
GTPγS	–	–	–	10 μL	100 μM
Antagonist	–	–	10 μL	–	10 μM
Membranes	165 μL	165 μL	165 μL	165 μL	11–15 μg
15′ incubation at 30 °C and gentle shaking at 400 rpm					
Agonist	–	10 μL	10 μL	–	10 μM
30′ incubation at 30 °C and gentle shaking at 400 rpm					
[^{35}S]GTPγS	5 μL	5 μL	5 μL	5 μL	0.4 nM
90′ incubation at 30 °C and gentle shaking at 400 rpm					
Detergent solution	20 μL	20 μL	20 μL	20 μL	
30′ incubation at 22 °C and gentle shaking at 350 rpm					
Antibody (in BSA 15%)	10 μL	10 μL	10 μL	10 μL	
90′ incubation at 22 °C and gentle shaking at 350 rpm					
PVT protein A SPA beads	50 μL	50 μL	50 μL	50 μL	0.75 mg
3 h incubation at 22 °C and gentle shaking at 350 rpm					
Seal plate					
Centrifugation for 8′ at 3,000 rpm and RT/quantification					

without destroying the [^{35}S]GTPγS-Gα association and to avoid the effect of proteases that could reduce the signal over time [17].

7. Incubation for 30 min at 22 °C and gentle orbital shaking.
8. After solubilization, 10 μL of diluted selective antibodies for each Gα subunit are added to each well (*see* **Note 5**). Antibodies are pre-diluted in 15% BSA at different concentrations for corresponding Gα-protein subtype (recommended antibodies and dilutions to be used are described in Table 1), and incubate for 90 min at 22 °C and gentle orbital shaking.
9. Finally, 50 μL of polyvinyltoluene (PVT) protein A-coated scintillation proximity assay (SPA) beads (0.75 mg/well) are added, and plates are incubated for 180 min at 22 °C with gentle agitation. Each batch of 500 mg of PVT protein A SPA (RPNQ0019) beads is previously resuspended in 33.5 mL of incubation buffer to leave it at a concentration of 15 mg/mL (*see* **Notes 3** and **4**).
10. Plates are sealed with PlateSeal (Perkin Elmer) and centrifuged 8′ at 1,000 × g and 22 °C with Eppendorf 5810R centrifuge and A-4-62 rotor.
11. Detection of bound radioactivity in a MicroBeta TriLux scintillation counter using 96-well rigid plate cassettes (Perkin Elmer: 1450-105) and MicroBeta TriLux program with Non-Paralux mode, half-life correction, and manual window: 5-600, for 1.5 min/well.

3.3 Data Analysis

Data are analyzed using GraphPad PrismTM 9.3.1 software.

Prior to data analysis, dpm values are transformed to fmol/mg protein. For this transformation, the following equation and parameters are used:

$$\text{Binding values (fmol/mg prot.)} = \text{binding values (dpm)} \bullet \frac{1\ \text{Ci}}{2.22\ 10^{12}\ \text{dpm}} \bullet \frac{\text{mmol}}{1{,}250\ \text{Ci}} \bullet \frac{10^{12}\ \text{fmol}}{\text{mmol}} \bullet \frac{1}{X\ \text{mg prot.}}$$

1 Ci = 2.22×10^{12} disintegrations per minute (dpm)

Specific activity of the radioligand ([^{35}S]GTPγS) = 1,250 Ci/mmol

X = amount of protein per well in mg

Then, the nonspecific binding value (NBS) must be subtracted from the [^{35}S]GTPγS binding data obtained for each condition. These specific values are represented as percentage of basal binding (BB), in the absence of compounds, for each Gα subunit. Thus, an agonist drug will increase the binding value with respect to baseline. In the case of the tests at a single concentration of the drugs, the statistical comparison between these values and 100%, considered the basal binding (BB) value, is carried out using Student's *t*-test of one sample. In the concentration-response curve experiments, the data are analyzed by nonlinear adjustment assuming a model of a binding site that allowed calculating the pharmacological parameters of maximum stimulatory or inhibitory effect (E_{max}/I_{max}) and the concentration that allows reaching 50% of that maximum stimulatory or inhibitory effect (EC_{50}/IC_{50}).

4 Notes

1. Membrane preparation should not be reused and each aliquot should be used just once. This requires a preliminary calculation of the following assays in order to determine the best amount of protein per aliquot.
2. Samples stored at −80 °C have no deterioration even after at least 1 year.
3. When protein A-coated SPA beads are used, the different antibody-capture affinity of this protein depending on the species of primary antibody should be considered when selecting the anti-G-protein antibody. In particular, protein A shows high binding affinity for rabbit IgG and mouse IgG2a and 2b [12, 13].
4. Other coatings of PVT SPA beads could be considered if analyses of the whole population of G proteins are required (WGA coated), etc.
5. Some antibodies have cross-reactivity and might detect more than one Gα-protein subtype. A previous characterization of the antibody via Western blot assay is always necessary to ensure the selectivity of the antibody. Every time we start using a different antibody, we should characterize also the concentration that will be used for the experiment. Optimization of experimental conditions is necessary to maximize the detection ratio.
6. To test the selectivity of a drug in the presence of a selective antagonist, we usually used one-point concentration experiments instead of concentration-response curves choosing that concentration in base of occupancy of the receptor by the tested drug and the antagonist, usually 1–10 μM of the antagonist vs 10 μM of drug:

   $$\% \text{ of occupancy (one drug)} = [D]/([D] + Kd)$$

   $$\% \text{ of occupancy (competition A vs B)} = [DA]/(([DA] + KdA(1 + ([DB]/KdB))$$

7. The experiments are performed usually in triplicate for each concentration point, in duplicates for nonspecific binding determination, and in quadruplicate for basal binding

measurements. To draw a concentration-response curve, a minimum of eight different concentration points should be tested.

8. The optimal membrane concentration should be determined preliminary depending on the receptor of study and the species (human, mouse, etc.) and sample source (brain, cell line, etc.).
9. Recommended GDP concentrations for the different Gα subunit subtypes: 100 μM for $G\alpha_{i1}$, $G\alpha_{i3}$, $G\alpha_z$, $G\alpha_s$ and $G\alpha_{12/13}$; 50 μM for $G\alpha_{i2}$ and $G\alpha_o$, and 50–10 μM for $G\alpha_{q/11}$. Stock $2\ 10^{-2}$ M GDP aliquots stored at −20 °C are used for freshly prepared necessary dilutions of GDP.
10. As concentration-response curves are really expensive for these assays, we usually use 10 μM single concentration for preliminary profile description. This concentration was selected because it gives binding values around maximal effects in the concentration-response curves.
11. As described in Table 1 for GDP, GTPγS and drugs are 10 μL pipetted in a final incubation volume of 200 μL, that is, diluting the concentration 20x times. This dilution factor has to be taken into account previously when stock solutions are prepared in the suitable solution (dimethylsulfoxide (DMSO), EtOH, incubation buffer, H_2O, etc.) depending on the solubility properties of the drug.
12. Stock solutions of drugs (usually $2\ 10^{-2}$ M) are prepared in the suitable solution (dimethylsulfoxide (DMSO), EtOH, incubation buffer, H_2O, etc.) depending on the solubility properties of each drug. This solution would be considered as the vehicle and should be present in the same proportion in all the wells. A vehicle of DMSO:incubation buffer in 1:1 proportion has been demonstrated to be useful in most cases and not to affect results and reproducibility of the assay.
13. In Table 2 particular case, both agonist and antagonist are dissolved in the same vehicle. If this is not the case, recalculation of volumes should be carried out, usually, reducing membrane volume but keeping the final amount of protein per well.

Acknowledgments

This work was supported by the Basque Government (KK-2019/00-49) to RD-A. IM-A was a recipient of a predoctoral fellowship from the Basque Government.
The authors would like to thank the staff members of the Basque Institute of Legal Medicine for their cooperation.

References

1. Alexander SPH, Christopolus A, Davenport AP et al (2019) The concise guide to pharmacology 21/22: G protein-coupled receptors. Br J Pharmacol 1:S27–S156. https://doi.org/10.1111/bph.15538
2. Sriram K, Inser PA (2018) Protein-coupled receptors as targets for approved drugs: how many targets and how many drugs? Mol Pharmacol 93(4):251–258. https://doi.org/10.1124/mol.117.111062
3. Syrovatkina V, Alegre KO, Dey R et al (2016) Regulation, signaling, and physiological functions of G-proteins. J Mol Biol 428(19): 3850–3868. https://doi.org/10.1016/j.jmb.2016.08.002
4. Marinissen MJ, Gutkind JS (2001) G-protein-coupled receptors and signaling networks: emerging paradigms. Trends Pharmacol Sci 22(7):368–376. https://doi.org/10.1016/s0165-6147(00)01678-3
5. Milligan G, Kostenis E (2006) Heterotrimeric G-proteins: a short history. Br J Pharmacol 147 (Suppl 1):S46–S55. https://doi.org/10.1038/sj.bjp.0706405
6. Hilger D, Masureel M, Kobilka BK (2018) Structure and dynamics of GPCR signaling complexes. Nat Struct Mol Biol 25(1):4–12. https://doi.org/10.1038/s41594-017-0011-7
7. González-Maeso J, Rodríguez-Puertas R, Gabilondo AM et al (2000) Characterization of receptor-mediated [^{35}S]GTPγS binding to cortical membranes from postmortem human brain. Eur J Pharmacol 390(1–2):25–36. https://doi.org/10.1016/s0014-2999(99)00827-4
8. Harrison C, Traynor JR (2003) The [^{35}S]GTPγS binding assay: approaches and applications in pharmacology. Life Sci 74(4): 489–508. https://doi.org/10.1016/j.lfs.2003.07.005
9. Seifert R, Wenzel-Seifert K (2002) Constitutive activity of G-protein-coupled receptors: cause of disease and common property of wild-type receptors. Naunyn Schmiedeberg's Arch Pharmacol 366(5):381–416. https://doi.org/10.1007/s00210-002-0588-0
10. Diez-Alarcia R, Odagaki Y, Miranda-Azpiazu P et al (2021) Functional approaches to the study of G-protein-coupled receptors in postmortem brain tissue: [35S]GTPγS binding assays combined with immunoprecipitation. Pharmacol Rep 73(4):1079–1095. https://doi.org/10.1007/s43440-021-00253-z
11. DeLapp NW (2003) The antibody-capture [^{35}S]GTPγS scintillation proximity assay: a powerful emerging technique for analysis of GPCR pharmacology. Trends Pharmacol Sci 25(8):400–401. https://doi.org/10.1016/j.tips.2004.06.003
12. DeLapp NW, McKinzie JH, Sawyer BD et al (1999) Determination of [^{35}S]GTPγS guanosine-5′-*O*-(*3*-thio)triphosphate binding mediated by cholinergic muscarinic receptors in membranes from chinese hamster ovary cells and rat striatum using an anti-G protein scintillation proximity assay. J Pharmacol Exp Ther 289(2):946–955
13. Richman DD, Cleveland PH, Oxman MN et al (1982) The binding of staphylococcal protein A by the sera of different animal species. J Immunol 128(5):2300–2305
14. Frank MB (1997) Antibody binding to protein A and protein G beads. Molecular Biology Protocols, Oklahoma City
15. Cook N, Harris A, Hopkins A et al (2002) Scintillation proximity assay (SPA) technology to study biomolecular interactions. Curr Protoc Protein Sci 27(1):19–18. https://doi.org/10.1002/0471140864.ps1908s27
16. Berry J, Price-Jones M, Killian B (2012) Use of scintillation proximity assay to measure radioligand binding to immobilized receptors without separation of bound from free ligand. Methods Mol Biol 897:79–94. https://doi.org/10.1007/978-1-61779-909-9_4
17. Harder D, Fotiadis D (2012) Measuring substrate binding and affinity of purified membrane transport proteins using the scintillation proximity assay. Nat Protoc 7(9):1569–1578. https://doi.org/10.1038/nprot.2012.090

Chapter 4

CRISPR/Cas9 as a Simple Technique for the Generation of Murine Knockout Models for Neuropsychiatric Diseases

Lide Totorikaguena and Estibaliz Olabarrieta

Abstract

The development of schizophrenia-like rodent models is still a major challenge for the study of this mental disorder. Schizophrenia and other neuropsychiatric disorders are thought to be triggered by multiple factors, and furthermore, the genetic component of schizophrenia is highly complex. The edition of one single gene for mimicking some of the symptoms of the disorder could cause unintended mutations that could influence animal's behavior making it difficult to study. Since 2013, CRISPR-Cas gene-editing technology has been a great improvement in the specificity of transgenic model generation because of its speed, efficiency, cost, and apparent ease. This protocol describes a simple method to generate a knockout mouse model using CRISPR technology, which can be applied to any gene presumably involved in the development of schizophrenia and other neuropsychiatric disorders.

Key words Mice, Gene edition, CRISPR, Cas9, Schizophrenia

1 Introduction

Mental illnesses, such as schizophrenia, can be triggered by multiple factors. The development of therapeutic drugs for this disease relies on the understanding of the basic biology and pathophysiology by the study of in vivo models [1]. Unfortunately, working with animal models of mental illness, inherent to human, is difficult as they do not fully reflect what happens in the human brain. To date, here are various types of animal models of schizophrenia: pharmacological models, genetic models, and neurodevelopmental models [2].

With regard to genetic models, it must been taken into account that the genetic component of schizophrenia is highly complex and probably there are various genes implicated. To date, there are various genetically modified mice to evaluate schizophrenia (e.g., DISC1, NRG1, erbB4, Munc 18-1, dysbindin) [2, 3]. However,

Leyre Urigüen and Rebeca Diez-Alarcia (eds.), *Schizophrenia: Methods and Protocols*, Methods in Molecular Biology, vol. 2687,
https://doi.org/10.1007/978-1-0716-3307-6_4,

edition of one gene could cause unintended mutations that could influence animal's behavior. Furthermore, generation of genetic models is a complicated long process and very expensive.

In recent years CRISPR-Cas gene-editing technology has been a great improvement in the specificity of transgenic model generation. Because of its speed, efficiency, cost, and apparent ease, it has been a true revolution in the field [4]. CRISPR-Cas has two main components: Cas enzyme (an RNA-guided endonuclease) and the RNA guide [5]. Cas enzyme requires an RNA molecule (the sgRNA) to guide the endonuclease to a specific location in the genome (known as gene target). sgRNA contains two different components: the crRNA (approximately 20 nucleotides that are complementary to the target DNA sequence) and the tracrRNA (RNA fragment that will bind to the endonuclease and create the ribonucleoprotein complex, RNP). Downstream to the complementary sequence in the target DNA, the protospacer adjacent motif (PAM) is located, which will be the one that recognizes the Cas endonuclease. So, both sgRNA and PAM are required for the endonuclease to bind the target genomic sequence and generate a double-strand break (DSB) on the DNA (Fig. 1).

When CRISPR system is introduced in a mouse zygote, different outcomes are possible after DSB. The cell will repair the DNA damage by two different mechanisms: Non-homologous end joining (NHEJ) or homology-directed repair (HDR). The first one occurred when the cell does not have any homologous template to repair the damage, so the reparation can lead to nucleotide insertions, deletions, or modifications. The second one needs a donor homologous strain to repair the DSB. This method is useful when the animal model is a knock-in, as it is possible to introduce a specific sequence as donor strain. The main disadvantage of this technique is the effect on the off-targets. As the sgRNA is relatively

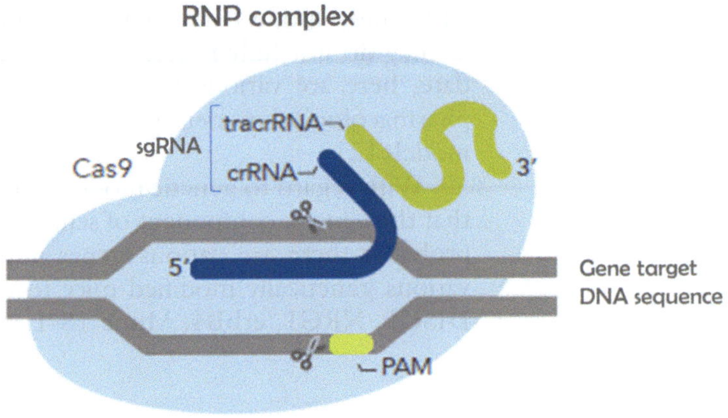

Fig. 1 CRISPR-Cas system components. (Modified from IDT guide for CRISPR edition)

short, there could be unspecific sites in the genome where the RNP could bind, so it is really important a proper design of the entire CRISPR system before starting the generation of a new mouse model [6, 7].

Finally, despite great advances in the generation of genetically engineered animal models, developing schizophrenia models is still a major challenge for scientists, for the reasons previously described. This protocol describes a simple method for the generation of a knockout mouse model using CRISPR technology, which can be applied to any gene involved in schizophrenia.

2 Materials

2.1 Animals

Female and male C57BL6/J and CD1 mice housed (aged 6–10 weeks) under controlled temperature (23 ± 1 °C), on a normal 12-h light/dark cycle, with free access to food and water.

2.2 Hormones and Drugs

- Equine chorionic gonadotropin (eCG) (Foligon, MSD, Animal Health, EEUU). Intraperitoneal administration, 5 IU
- Human chorionic gonadotropin (HCG) (Veterin Corion, Divasa-Farmavic, Spain). Intraperitoneal administration, 5 IU
- Buprenorphine (Ecuphar, Spain). Intraperitoneal administration before surgery (100 μg/kg)

2.3 TE Buffer

1. 10 mM Trizma base (Sigma Aldrich, T1503). Dissolve in 8 mL of H_2O 12.11 mg of Trizma Base.
2. 1 mM EDTA (Sigma Aldrich, E5134). Add 2.92 mg of EDTA.
3. Adjust to 7.5 pH with HCl.
4. Adjust volume to 10 mL.
5. Sterilize by filtration with 0.2 μm syringe filters (Merch Millipore, SLGP033RS). Store at 4 °C up to 1 year.

2.4 M2 Medium

Commercial or handmade medium are suitable for embryo handling (e.g., Sigma-Aldrich, M7167 or similar). It is important to check if medium need to be pH calibrated in CO_2 incubator before using or not. If that is the case, incubate M2 medium at least 2 h before starting experiment. M2 medium must be sterilized.

2.5 KSOM Medium

Commercial or handmade medium are suitable for embryo handling (e.g., Merck, MR-121 or similar). It is important to check if medium need to be pH calibrated in CO_2 incubator before using or not. If that is the case, incubate KSOM medium at least 2 h before starting experiment. KSOM medium must be sterilized.

2.6 STES Buffer

1. 50 mM Trizma base (Sigma Aldrich, T1503). Dissolve 0.302 g of Trizma base in 40 mL of H_2O.
2. 20 Mm NaCl (Sigma Aldrich, S5886). Add 0.058 g of NaCl.
3. 1 mM EDTA (Sigma Aldrich, E5134). Add 0.018 g of EDTA.
4. 0.1% SDS (Sigma Aldrich, L3771). Add 0.5 g of SDS.
5. Adjust to 8 pH with HCl.
6. Adjust volume to 50 mL.
7. Sterilize by filtration with 0.2 μm syringe filters. Store at 4 °C until 1 year.

2.7 LB Agar

1. Add 40 g of LB agar (Condalab, 1552.00) in 1 L of distilled water. Mix well and dissolve by heating with agitation.
2. Add 25 μg/mL of ampicillin (Fisher BioReagents, BP1760-25) and mix well.
3. Sterilize in autoclave at 121 °C for 15 min.

2.8 LB Broth

1. Add 25 g of LB broth (Condalab, 1551.00) in 1 L of distilled water. Mix well and dissolve by heating with agitation.
2. Sterilize in autoclave at 121 °C for 15 min.

3 Methods

All procedures with embryos should be carried out at 37 °C. CRISPR-Cas9 components should be maintained as cool as possible during performance. Workflow of entire experiment is shown in Fig. 2.

3.1 Designing RNA Guide for Your Gene Disruption with CRISPR

There are several bioinformatics tools for sgRNA designing (e.g., Benchling, Breaking-Cas). The optimal sgRNA should be selected according to the minimal distance from the gene target. It is recommended that the RNA guide have no more than 20–30 nucleotides for better specificity with the sequence target [8].

1. Look for the specific gene information in your chosen database (NCBI, Ensembl). Select the exon where the gene edition is preferred (*see* **Note 1**).

Fig. 2 Workflow in a knockout model generation by CRISPR-Cas

2. Select that region and create your guide. Choose the guide whose position is the closest to your intended mutation. It is recommended to verify the off-targets of your chosen sgRNA and select that one with more specificity.

3.2 Microinjection of Fertilized Mouse Oocytes

The following methods are based on previously described at Behringer et al. (2014) [9]. This procedure should start at 10 am to ensure that all ovulated oocytes are in the ampulla.

1. Sacrifice by cervical dislocation the female oocyte donors with a plug previously seen that morning.

   Embryo donors strain should be chosen depending on the background that it is needed for the knockout mouse model. Female donors must be superovulated to induce estral cycle and produce a larger number of oocytes and then mate with males to get the desire embryos (see **Note 2**). *For that purpose, intraperitoneal administration of 5 IU eCG 48 h before mating is needed. Then, just before preparing the natural mates, inject intraperitoneal 5 UI of HCG.*

2. Extraction of the cumulus oocytes complex (COC). Open the abdominal cavity and remove the ovaries attached to part of the uterus. Transfer them to a petri dish with M2 medium and extract mature oocytes from the ampulla with the aid of a needle.
3. Incubate oocytes 5 min in the presence of 0.3 mg/mL hyaluronidase (Sigma Aldrich, H3506) to remove granulose cells and wash oocytes in M2 medium.
4. Transfer the oocytes to a petri dish with KSOM medium and then to the culture dish with the same medium. Incubate 4 h at 37 °C, 5% CO_2, and 95% humidity.
5. Prepare microinjection set up during the oocyte incubation time (Fig. 3). Be sure to have proper pipettes for embryo handling and microinjection (Sutter Instrument Co. USA, Model P-97, Flaming/Brown micropipette puller). Power on the heating plate at 37 °C 15 min before starting the microinjection session and warm a petri dish with M2 medium. Prepare a petri dish with a M2 drop of 100 μL and cover it with mineral oil (Sigma Aldrich, M8410).
6. Prepare CRISPR-Cas9 reagents for microinjection. Add in a 0.2 mL tube 40 ng/μL of Cas9 (Takara, 632641) and 25 ng/μL of sgRNA and TE buffer to a total volume of 20 μL. Spin down the mixture and place it in ice until needed.
7. Perform microinjection in groups of 20 oocytes. Place a group in the M2 drop and microinject the CRISPR reagents into the cytoplasm of each oocyte. Microinjection can also be performed into the pronucleus (*see* **Note 3**) (Fig. 3a–c).
8. Remove microinjected oocytes and place them in M2 at 37 °C for 5 min until recovered.

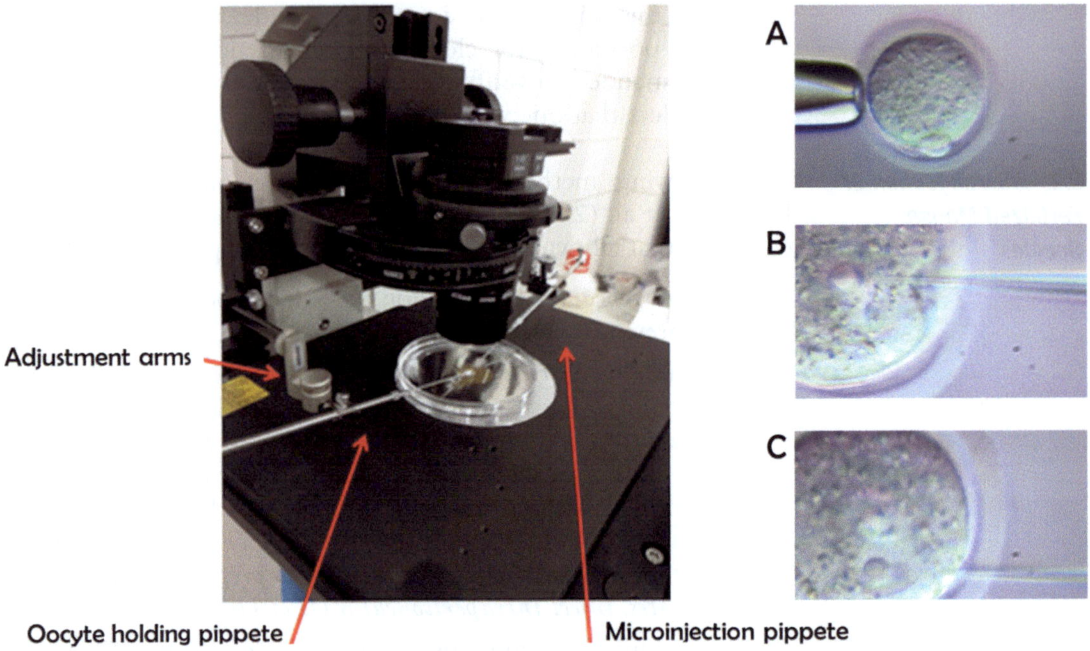

Fig. 3 Microinjection set (Nikon Eclipse Ti-S). (**a**), (**b**), and (**c**) represent steps for zygote pronuclear microinjection

3.3 Embryo Culture

1. Prepare culture dish with eight KSOM drops of 25 μL. Cover it with mineral oil for embryo culture.
2. After oocyte recover in M2 medium, wash them in KSOM and place them in the culture dish for 19 h until embryo transfer.

3.4 Preparing Female Recipient for Embryo Transfer

CD1 female mice (6–10 weeks) are commonly used as female recipients for embryo transfer. Pseudopregnant mice are prepared by mating females with vasectomized males [9]. In this protocol it is described the oviduct transfer, but it also can be performed in the uterus.

1. After microinjection, prepare female recipient for embryo transfer. Select females in estrus and place one female in a cage with one vasectomized male (prepare a total of ten natural matings).

   Male vasectomy. CD1 males should be vasectomized at least 2 weeks before being used for matings. Anesthetize with isoflurane and administrate intraperitoneal buprenorphine. Place the animal lying on his back and make a transverse incision on the skin to access testis. Grasp and gently pull out one of the testicular fats. The testicles, vas deferens, and epididymis will come with the fat pad. Cauterize the vas deferens and place back in the interior of the abdominal cavity. Perform the same for the second testicle. The abdominal wall will be sutured with two or three stitches. The skin will then be sutured or stapled (*see* **Note 4**).

3.5 Embryo Transfer

1. 17 h after microinjection (next day at 8 am), check the female plugs and select those which will be the recipients.
2. Take the culture embryo and count how many have divided into two cells. Place them in M2 medium and reserve until needed.
3. At 10 am, anesthetize the female mouse with isoflurane. Make a single and small incision in the skin to access the abdomen. The left ovary can be reached through this incision (*see* **Note 5**).
4. Using blunt forceps grasp and gently pull out one of the ovarian fat pads. The associated ovary, oviduct, and left uterine horn will come with the fat pad. Hold the fat pad with blunt forceps.
5. Locate the infundibulum and break the ovarian bursa. Using a pipette, take the embryos previously in M2 and introduce them through the infundibulum (*see* **Note 6**).
6. Carefully, place the ovary inside the abdominal cavity. The abdominal wall will be sutured with two or three stitches. The skin will then be sutured or stapled (*see* **Note 7**).

3.6 Genotyping

After 19 days of pregnancy, it will be expected the outcome of the first litter. At the age of 8 days, DNA samples are prepared from mouse tail section. The procedure to isolate DNA is performed by Proteinase K digestion. The genotyping and allele sequencing methods are based on those developed by the Department of Animal Reproduction of the National Food and Agriculture Research and Technology Institute (INIA, Spain).

1. Cut approximately 2 mm of mouse tail and place the tissue into a 1.5 mL microcentrifuge tube. Make sure that all mice are correctly identified. Tail tissue can be stored at −20 °C until DNA isolation.
2. Add 20 μL of STES buffer and 2.5 μL of 20 mg/mL Proteinase K solution (Macherey-Nagel, Germany).
3. Incubate at 55 °C for 2 h in constant agitation.
4. Add 80 μL of sterile H_2O and vortex.
5. Incubate the mixture at 25 °C for 10 min.
6. Add 200 μL of sterile H_2O and vortex.
7. Centrifuge at maximum speed for 3 min.
8. Prepare an aliquot of supernatant for PCR. If samples are not going to be immediately used, they can be stored at −20 °C.

9. For genotyping PCR specific primers for the sgRNA region should be used. PCR design depends on the characteristics of the chosen taq polimerase (consult the supplier's specifications). Final elongation of 30 min should be performed at 72 °C to ensure the correct performance of the following cloning steps (*see* **Note 8**).

   sgRNA primers can be designed with multiple tools (e.g., *BLAST primer design*). *It is recommended to select 60 bp upstream and 60 bp downstream the sgRNA designed. Thus, the genotyping PCR product will be approximately 140 bp. Bigger or smaller products are not recommended as it will be difficult to distinguish small base modifications.*

10. Prepare 2–3% agarose gel and load 10 μL of each sample. Run during 1 h at 120 V (*see* **Note 9**). Those samples in which a band size change is observed in comparison to the wild-type band (possible homozygote) or those in which two bands appear (possible heterozygote) should be selected for allele sequencing.

3.7 Allele Sequencing

Each sample selected corresponds to one possible modified animal. PCR results do not allow the study of the DNA base modification. It is necessary to sequence the alleles and evaluate the specific nucleotide modification in the genome region for the studied gene. It is highly recommended to repeat the PCR from selected samples so as to produce more PCR product for cloning steps (at least a total of 100 μL of PCR product).

1. Purify PCR products from selected samples using Favorprep Gel/PCR Purification Kit (Favorgen, FAGCK 001-1) following manufacturer's instructions. Elute in 40 μL and quantify in nanodrop the DNA concentration. A concentration between 20 and 40 ng/μL will be suitable for cloning steps.
2. Ligate the PCR product with pMD20 vector (Fig. 4; Takara Bio, Japan). Add in a 1.5 mL tube 0.5 μL of pDM20 vector, 4.5 μL of purified PCR product, and 5 μL of Blunt/TA ligase master mix (BioLabs, M0367L). Vortex the mixture and incubate at 25 °C for 15 min.
3. Transform the ligation mixture in DH5α competent cells (Fisher Scientific, 18265017). Add 3 μL of ligation mixture to 100 μL of competent cells. Be sure that all material is correctly clean and work under sterile conditions.
4. Incubate in ice for 30 min.
5. Immediately transfer to 42 °C for 30 s.
6. Maintain 2 min in ice.
7. Immediately after, add 900 μL of LB medium to the reaction mix and incubate at 37 °C for 1 h in constant agitation.

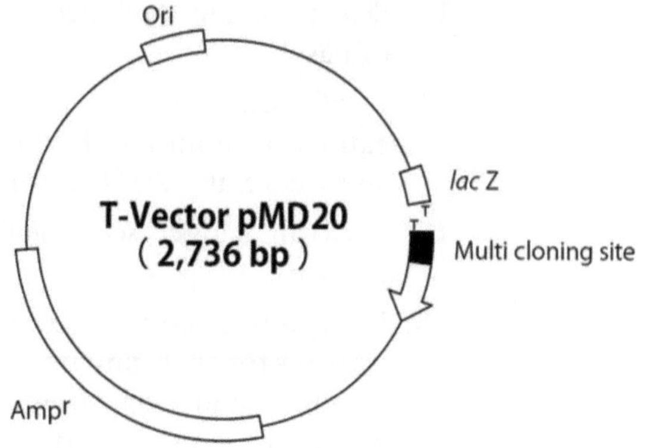

Fig. 4 pMD20-T vector (2736 bp). (Takara Bio Ref. 3270)

8. Centrifuge for 7 min at 1,500× *g* and discard supernatant.
9. Resuspend the pellet in the remaining medium and spread into a previously prepared LB plate.

   Preparation of LB plates: Dissolve LB agar following manufacturer's instructions. Add 25 μg/mL of ampicillin and throw 30 mL in each plate. Wait until the agar is solidified. 30 min before spread the transformation product in plates, add 50 μL of 20 mg/mL X-Gal (Invitrogen, 15520-034) and 150 μL of 0.1 M IPTG (Thermo Scientific, 34060) to each plate, and distribute over the entire plate.
10. Incubate LB plates 24 h at 37 °C.
11. Colony selection for sequencing. For better color identification, after incubation place the plates at 4 °C.

    As pMD20 vector present resistance to ampicillin, every DH5α colony grown in the plate with this antibiotic has been successfully transformed. Additionally, LB plates have been supplemented with X-galactosidase and IPTG. The LacZ genes encode β-galactosidase enzyme that hydrolyze X-galactosidase causing a blue color and IPTG induce β-galactosidase activity. pMD20 vector gene LacZ is located in the vector cloning site; thus, when successful cloning has been performed, LacZ will not produce a functional β-galactosidase and no blue color will be appreciated (white colonies).
12. Select 5–10 white colonies and inoculate in 3 mL LB. Incubate 1 h at 37 °C.
13. PCR of colonies. Take 10 μL from culture in a 0.2 mL tube. Add 10 μL of sterile H_2O and incubate 10 min at 95 °C.
14. Perform PCR according to the taq polimerase manufacturer's instructions. Use primers designed for the sgRNA or vector M13 primers (Takara Bio, Japan).

15. Select colonies in which PCR present a change in the number of bases.
16. Purify plasmids from those colonies with FavorPrep plasmid extraction mini kit (Favorgen, FAPDE 300). Elute in 50 μL and freeze at −20 °C until needed.
17. Perform Sanger sequencing to study the CRISPR genome edition.
18. Interpreting sequencing results. There are a lot of bioinformatics tools for analyzing nucleotide sequences (e.g., Bioedit). The selected sequences should be compared with the wild-type reference sequence. It should be noted that an effective sequence edition to generate a knockout mouse model needs to cause a change in the reading frame of the interest protein.

4 Notes

1. Designing sgRNA. A gene edition in the first exon (codifying region) will be more effective as the reading frame of the protein will be interrupt sooner, increasing the likelihood of a successful disappearance of the protein's functionality.
2. Classically, C57BL6/J strain is used for embryo donors in gene edition. In this strain, after superovulating 4 female mice, we will get approximately 80 oocytes (20 oocytes per female). The number of female donors should be chosen depending on the ability to microinject in short time, as it is not recommended to perform long sessions. Take into account that only 50–60% of microinjected oocytes will be suitable for embryo transfer. If other strain is going to be used, check before the reproductive characteristics.
3. If Cas9 protein is directly injected in the zygote, cytoplasm injection can be performed. If other Cas9 forms are chosen, revise the CRISPR system deliver method or inject directly in one of the pronucleus. Pronuclear microinjection involves increase practice from the user.
4. It is recommended to check the correct vasectomy of males before starting the experiment. For that purpose, mate the vasectomized males with 2–3 females and check the production of plugs and no pregnancy.
5. For unilateral embryo transfer, a maximum of 20 microinjected embryos should be transferred. Bilateral embryo transfer can be performed, but in this case, transfer only a maximum of ten oocytes per oviduct.
6. When introducing the embryos through the infundibulum, make sure there is no reflux and they stay inside the oviduct. To do this, use absorbent material to remove excess fluid from

the broken bursa or blood from broken capillaries. If you load the pipette with a small bubble at the end of the embryos, it will be easier to see that you have introduced the embryos, as you will see the bubble entering the infundibulum.

7. When transferring the mouse to the cage, do it in a horizontal position so that there are no sudden movements and the embryos are placed in the oviduct. Too much movement decreases the chances of a successful transfer. Furthermore, house the female mice in a room without males during the pregnancy to avoid the Bruce effect.
8. T-Vector pMD20 vector is linearized and the T-overhang ends at the cloning site improve the efficiency of ligation of PCR products that contain A-overhangs at 3′-ends (Takara, Ref. 3270). The final elongation of 30 min allowed the PCR product to bind the vector in the ligation step.
9. If band sizes are not properly differed, increase the percentage of agarose or run the gel during more time at lower intensity to better distinguish the difference between bands, as sometimes the number of nucleotides changes is too low.

Financial Support

This work was supported by the Basque Government (IT1512–22).

References

1. Wang P, Li M, Zhao A, Ma J (2021) Application of animal experimental models in the research of schizophrenia. Am J Med Genet B Neuropsychiatr Genet 186(4):209–227
2. Bialon M, Wasik A (2022) Advantages and limitations of animal schizophrenia models. Int J Mol Sci 23:5968
3. Mollinedo I (2017) In vivo functional role of the 5-HT2A/mGlu2 receptor heterocomplex in rodents. Doctoral thesis
4. Li C, Du Y, Zhang T, Wang H, Hou Z, Zhang Y, Cui Y, Chen W (2022) "Genetic scissors" CRISPR/Cas9 genome editing cutting-edge biocarrier technology for bone and cartilage repair. Bioact Mater 22:254–273
5. Zhang F, Wen Y, Guo X (2014) CRISPR/Cas9 for genome editing: progress, implications and challenges. Hum Mol Genet 23(R1):R40–R46
6. Mojica FJM, Montoliu L (2016) On the origin of CRISPR-Cas technology: from prokaryotes to mammals. Trends Microbiol 24:811–820
7. Sander JD, Joung JK (2014) CRISPR-Cas systems for genome editing, regulation and targeting. Nat Biotechnol 32:347–355
8. Fernández A, Morín M, Muñoz-Santos D, Josa S, Montero A, Rubio-Fernández M, Cantero M, Fernández J, Del Hierro MJ, Castrillo M et al (2020) Simple protocol for generating and genotyping genome-edited mice with CRISPR-Cas9 reagents. Curr Protoc Mouse Biol 10:e69
9. Behringer R (2014) Manipulating the mouse embryo: a laboratory manual. Cold Spring Harbor Laboratory Press

Chapter 5

Evaluation of Sensorimotor Gating Deficits in Mice Through Prepulse Inhibition (PPI) of the Startle Response

Paula Unzueta-Larrinaga and Leyre Urigüen

Abstract

Prepulse inhibition of the startle response enables measuring animal behavior and helps us understand core aspects of neuropsychiatric diseases. Prepulse inhibition is considered a translational indicator of sensorimotor gating deficits present in schizophrenia patients and is crucial in the characterization of animal models of schizophrenia-like behaviors. Hallucinogenic drugs acting through 5-HT2A receptors, such as psilocybin, lysergic acid diethylamide (LSD), and dimethoxyiodoamphetamine (DOI), produce symptoms in healthy subjects comparable to those seen in schizophrenia and can be used in rodent models for mimicking some of these behaviors. Here we describe a protocol for the evaluation of prepulse inhibition of the startle response in CD1-Swiss male mice after a single dose of the hallucinogenic drug DOI.

Key words Prepulse inhibition, Animal models, Psychiatric disorders, DOI

1 Introduction

Deficits in the prepulse inhibition (PPI) of the startle response have been the focus of intense study in schizophrenia in both humans [1] and animal models [2, 3]. The startle response is a brainstem reflex response to sudden unexpected or threatening acoustic or tactile stimulus and is associated with negative affect. The prepulse inhibition evaluates the innate reduction of the startle reflex induced by a weak prestimulus.

Nowadays, PPI test has made a tremendous impact in the field of behavioral tasks for the characterization of the antipsychotic effects [4] of developing drugs and the exploration of the mechanisms underlying psychotic-like behaviors [5, 6].

A deficit in the PPI has been observed in psychiatric patients, especially in those with a diagnosis of schizophrenia [7]. Rodent models resembling some of the symptoms of psychosis or schizophrenia display impaired sensorimotor gating that can be quantified by the PPI of startle test. These models have demonstrated a clear

Leyre Urigüen and Rebeca Diez-Alarcia (eds.), *Schizophrenia: Methods and Protocols*, Methods in Molecular Biology, vol. 2687,
https://doi.org/10.1007/978-1-0716-3307-6_5,

face, predictive, and construct validity for schizophrenia therapeutics, neurophysiological endophenotypes, and potential causative insults for this group of disorders [8].

While there are considerable challenges in the translation of the neural substrates regulating PPI from human to rat/mouse, this test is being used successfully in a number of important lines of research, from the development of novel antipsychotic agents [9, 10] to the identification of vulnerability genes associated with schizophrenia [11, 12]. These efforts would benefit greatly from a map of the neural circuit regulation of PPI in humans, which could guide the search for functional targets of therapeutic drugs and candidate genes.

Hallucinogenic drugs acting as 5-HT2A receptor agonists, such as psilocybin, lysergic acid diethylamide (LSD), and dimethoxyiodoamphetamine (DOI), produce symptoms in healthy subjects comparable to those seen in schizophrenia and can be used in rodents for mimicking some of these behaviors. Here, we describe a protocol for the evaluation of PPI of the startle response in CD1-Swiss male mice after a single dose of the 5-HT2A hallucinogenic agonist DOI.

2 Materials

2.1 Animals

1. Male CD-1 mice housed (aged 6–8 weeks, 6–8 animals per cage) under controlled temperature (23 ± 1 °C), on a normal 12-h light/dark cycle (lights on between 8:00 and 20:00) and with free access to food and water.

2.2 Drugs and Administration Material

1. Syringe and needle for injection.
2. Ultra-pure water (mQ H_2O) (prepared by purifying deionized water, sensitivity of 18 MΩ-cm at 25 °C).
3. Saline solution: 0.9% NaCl. Weigh 0.9 g of NaCl and dissolve on 100 mL of mQ H_2O.
4. Dimethoxyiodoamphetamine (DOI) solution for intraperitoneal (i.p.) injection. Dissolve 1 mg of DOI in 10 mL of saline in order to get a 0.1 mg/mL solution.

This concentration was chosen in order to get a dose of 0.5 mg/kg, dose that has been previously characterized as the optimal for testing the PPI in this mice strain [13], when using 5 mL/kg as maximal i.p. injection volume. Thus, for injecting a mouse weighing 40 g, the i.p. injection volume of a 0.1 mg/mL solution of DOI will be 0.2 mL to get a dose of 0.5 mg/kg.

**The volume of injection is calculated according to the weight of each animal.*

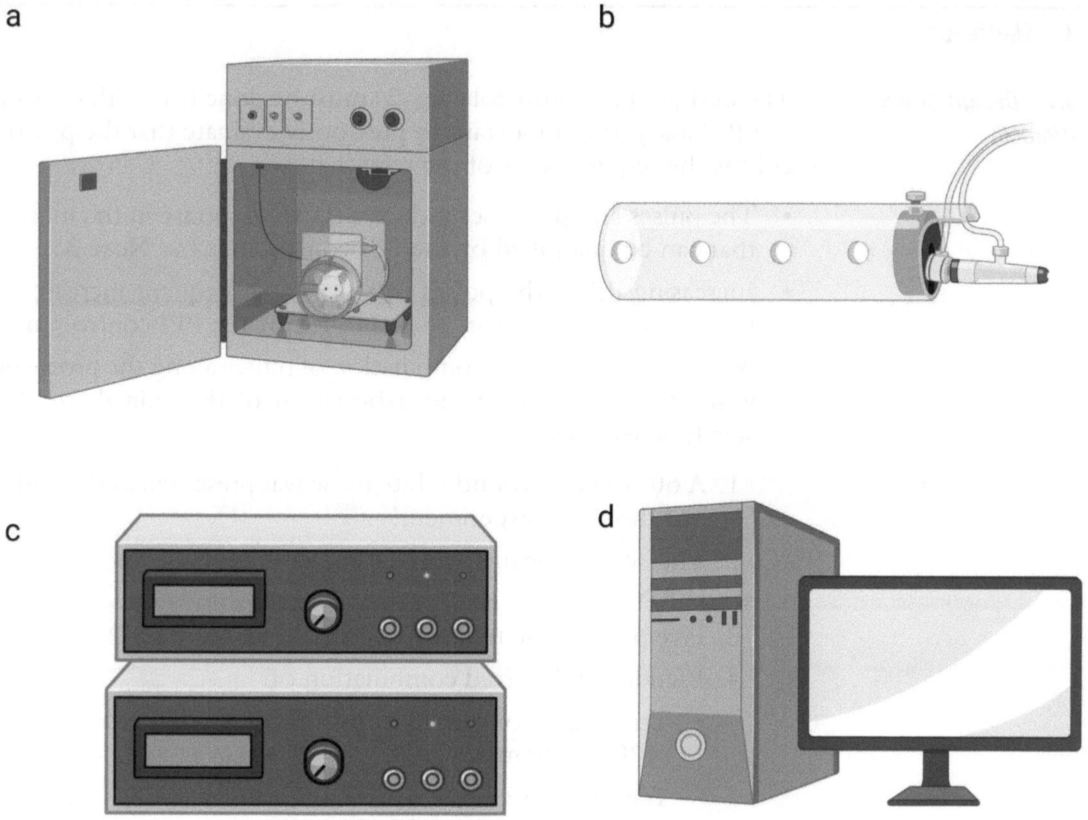

Fig. 1 Components of the prepulse inhibition test experimental equipment. Outer attenuated chamber (**a**), animal holder (**b**), control unit: transducer and amplifier, (**c**) and computer and appropriate software (**d**)

The final volume to be injected via *i.p. is always:*

$$\text{Volume} = \frac{\text{Weight (g)}}{2 \times 10^2}$$

2.3 Equipment

There are different suppliers for the PPI test equipment. Here is described the protocol for the *StartFear Combined System* from PanLab (Harvard Apparatus).

The experimental equipment (Fig. 1) consists of:

1. An attenuated chamber that serves to avoid external vibrations, noise, or other issues that may interfere with the experiment.
2. An animal holder (*see* **Note 1**).
3. A load cell platform inside the chamber that records the startle response of the animal.
4. A transducer and an amplifier linked to the load cell platform.
5. A sound generator and the appropriate software (e.g., PACK-WIN Software, PanLab, Harvard Apparatus) that regulates pulses and analyzes the results (*see* **Note 2**).

3 Methods

3.1 Design of the Protocol

The design of the protocol (Fig. 2) must be done before the day of the PPI assay. It is important to previously evaluate that the protocol fits the requirements of the test, that is:

- The pulses and prepulses induce a startle response in the mouse that can be quantified by the PPI control unit (*see* **Note 3**).
- Increasing dB of the pulses produces a subsequent increase in the startle reflex that can be quantified by the PPI control unit.
- A pseudo-randomized combination of pulses along the protocol is needed to minimize the habituation of the animals to the startle across trials.

1. A 60-dB background-white noise was presented as throughout the entire experiment.
2. The session begins with 10 min of acclimatization period (*see* **Note 4**).
3. Five pulse-alone trials of 120-dB white noise for 20 ms.
4. Pseudo-randomized combination of:
    - Ten pulse-alone trials consisted of a white noise burst (120 dB, 40 ms).
    - Ten prepulse-alone trials for each prepulse intensity (10 ms at 77, 82, or 87 dB).
    - Ten prepulse-pulse trials for each prepulse intensity with an interval of 60 ms between them and ten no-stimulus trials, in which only the background noise is present.

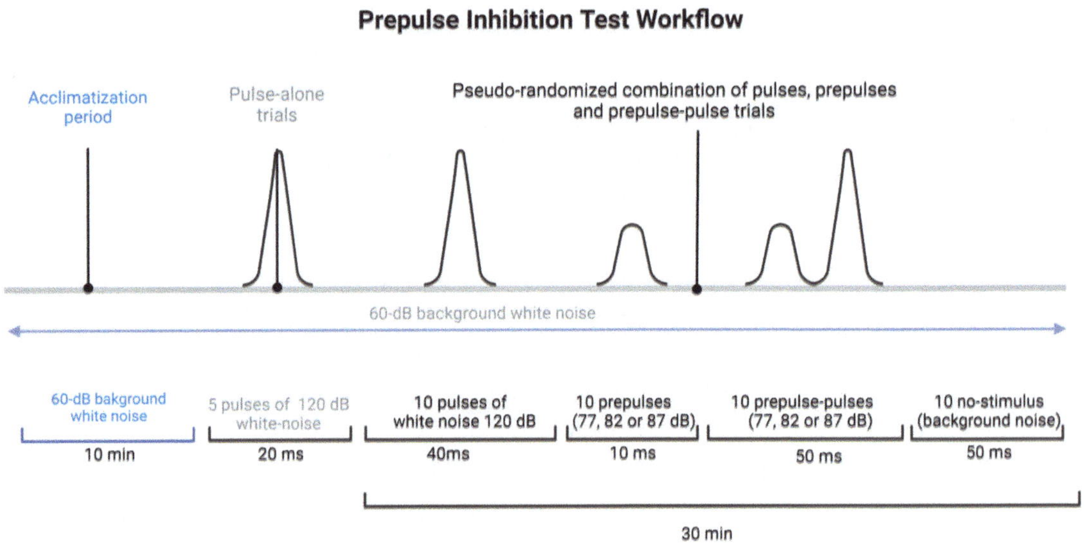

Fig. 2 Complete PPI protocol workflow

- The intertrial intervals were 10, 12, 15, 20, and 25 s. The trials are presented in a randomized order within each block, with the intertrial interval (ITI) varying randomly between 10 and 25 s, intended to minimize habituation to startle across trials (*see* **Note 5**).
- The maximum amplitude of the startle reaction is recorded for every trial, and the average startle for each trial subgroup is used for the analysis.
- No-stimulus trials and prepulse-alone trials do not exert any startle response and data are not included in the analysis.

3.2 Administration of DOI

1. Load the syringe with the desired volume of DOI or saline.
2. Carefully immobilize the animal.
3. Inject intraperitoneally (i.p.).
4. Leave the animal in the cage for half an hour before performing the PPI test.

3.3 Prepulse Inhibition Assay

1. Turn on the computer.
2. Switch on the control unit.
3. Switch on the box.
4. Open the software and choose the selected protocol.
5. Put the animal in the corresponding box and close the holder.
6. Start running the protocol.
7. When the first pulses come out, adjust the gain (Fig. 3). The pulse must be in range. If it is out of range, adjust the gain in the load cell coupler of the box (*see* **Note 6**).
8. At the end of the test, click to save the results, close the window, remove the animal from the chamber, and put it back on the cage (Table 1).
9. Wipe clean the animal holder and allow time to dry before loading another animal. The test has finished. Start again to carry out the next experiment (*see* **Note 7**).

3.4 Analyses of the Data

1. The maximal peak amplitude is used to determine the acoustic startle response.
2. PPI is calculated as a percentage score:
   % PPI = 100 – [(startle response to prepulse-pulse trial/ startle response to pulse-alone trial) × 100].
3. The appropriate PPI software (e.g., PACKWIN Software, Pan-Lab, Harvard Apparatus) is necessary for the analyses of the data (Fig. 4).

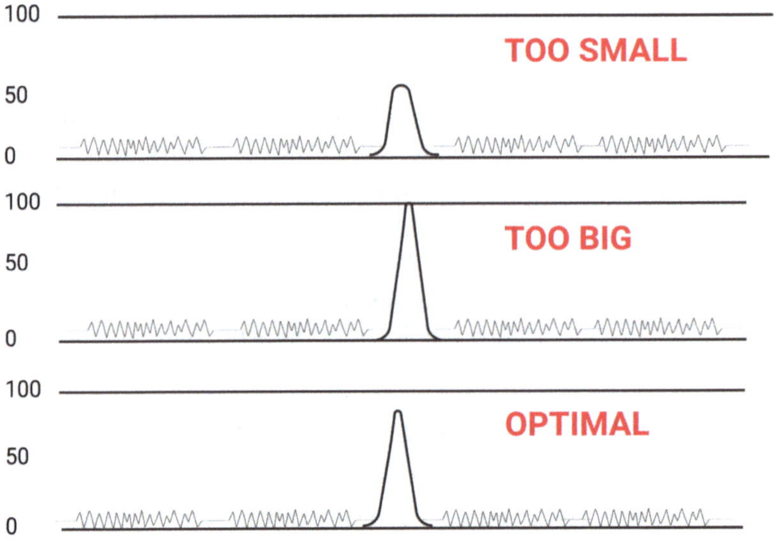

Fig. 3 Graphical representation of the gain. The gain is the parameter that controls the amplification for the activity signal received from the load cells

Table 1
Example of metadata obtained after a PPI session

State	Total Ent.	Mean Max.	Mean Lat. Max.	Mean Lat. Onset	Mean Duration	Mean Area	
Exploration	1	100,00	259889,00	259886,00	1589,00	45,85	
Trial Pulse Basal	6	10,50	49,17	0,00	0,00	0,00	
SR Pulse Basal	5	57,20	114,40	23,20	56,80	10,40	
ITI Basal	5	17,46	4021,00	0,00	0,00	0,00	
Trail No-STIM	10	12,29	49,70	0,00	0,00	0,00	
SR No-STIM	10	72,23	10309,70	875,70	106,90	10,29	
Trial Pulse	10	19,48	48,50	0,00	0,00	0,00	
SR Pulse	10	61,01	188,60	23,90	86,30	11,27	
Trial PrePulse 77 dB	10	15,57	54,60	0,00	0,00	0,00	
SR PrePulse 77 dB	10	19,97	250,00	0,00	0,00	0,00	
Trial PrePulse Pulse 77 dB	10	17,68	41,70	0,00	0,00	0,00	
PrePulse 77 dB	10	12,84	30,30	0,00	0,00	0,00	
SR PrePulse Pulse 77 dB	10	33,62	190,60	0,00	0,00	0,00	100-(33,62/61,01)x100=44,89
Trial PrePulse 82 dB	10	20,75	52,10	0,00	0,00	0,00	
SR PrePulse 82 dB	10	20,47	79,10	0,00	0,00	0,00	
Trial PrePulse Pulse 82 dB	10	11,28	30,50	0,00	0,00	0,00	
PrePulse 82 dB	10	11,86	36,60	0,00	0,00	0,00	
SR PrePulse Pulse 82 dB	10	23,62	201,50	0,00	0,00	0,00	100-(23,62/61,01)x100=61,28
Trial PrePulse 87 dB	10	15,53	35,20	0,00	0,00	0,00	
SR PrePulse 87 dB	10	20,64	193,70	0,00	0,00	0,00	
Trial PrePulse Pulse 87 dB	10	16,13	47,10	0,00	0,00	0,00	
PrePulse 87 dB	10	11,76	15,70	0,00	0,00	0,00	
SR PrePulse Pulse 87 dB	10	17,29	200,80	0,00	0,00	0,00	100-(17,29/61,01)x100=71,66
ITI	80	52,52	8285,55	1072,04	58,66	5,38	
Totals	287	31,63	3708,24	1236,10	29,61	2,59	

The table shows the results of a session with one mouse. The mean intensities of the startle response to the pulse and the prepulse-pulses at three different dB are highlighted in yellow. In the right column, these data are used to obtain the PPI value for each dB (77, 82, and 87) following the formula described in 3.3. section

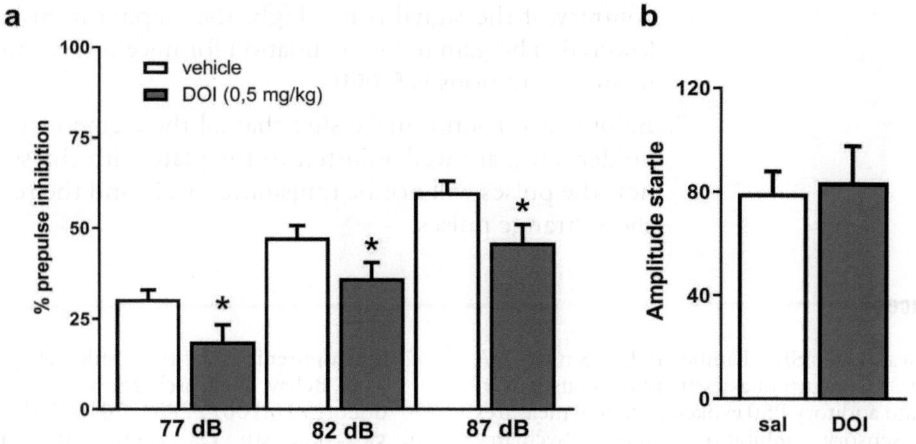

Fig. 4 Acute DOI disrupts the PPI. (**a**) DOI-induced PPI disruption in the three pulse intensities evaluated (student's *t*-test *$p < 0.05$) compared to saline. (**b**) No differences were found in the startle amplitude among groups. Bars represent mean ± SEM

4 Notes

1. When the mouse is small, it can escape from the holder even if it is tightly closed. To prevent this, cover the holder opening with a piece of adhesive tape.
2. Calibration of the sound and the movement sensors is very important for obtaining valid test results. Consequently, these must be routinely calibrated before each experiment. A commercial sound calibrator can be used for the calibration of the equipment.
3. It is important to note that the strain of the mouse and the genetic background, pharmacological treatment, age, or even physiological status may have an impact on the response in the PPI test. Therefore, it may be necessary to adjust the protocol (dB, duration of the pulses, etc.) depending on these different conditions.
4. This period is used for the animal to get used to the holder and the startle box and to keep it calm and without stress when performing the test.
5. When designing the protocol, it is important to take into account that ITIs below 10 s should be avoided in order to exclude effects caused by muscle fatigue and refractory periods of muscle responses.
6. The gain is a parameter that controls the amplification for the activity signal received from the load cells. If the signal is low, the amplification has to be manually increased, and on the

contrary, if the signal is too high, the amplification has to be lowered. The gain recommendation for mice under our experimental conditions is 5,000.

7. Before each round, make sure that all the elements (load cell, holder, etc.) are well adjusted to the plate with the sensors. If not, the pulses will not be transmitted well, and the record will show strange pulses.

References

1. Schwarzkopf SB, Lamberti JS, Smith DA (1993) Concurrent assessment of acoustic startle and auditory P50 evoked potential measures of sensory inhibition. Biol Psychiatry 33(11–12):815–828
2. Carter RJ, Lione LA, Humby T, Mangiarini L, Mahal A, Bates GP, Dunnett SB, Morton AJ (1999) Characterization of progressive motor deficits in mice transgenic for the human Huntington's disease mutation. J Neurosci 19(8): 3248–3257
3. Frankland PW, Wang Y, Rosner B, Shimizu T, Balleine BW, Dykens EM, Ornitz EM, Silva AJ (2004) Sensorimotor gating abnormalities in young males with fragile X syndrome and Fmr1-knockout mice. Mol Psychiatry 9(4): 417–425
4. Xue YY, Wang HN, Xue F, Tan QR (2012) Atypical antipsychotics do not reverse prepulse inhibition deficits in acutely psychotic schizophrenia. J Int Med Res 40(4):1467–1475
5. Geyer MA (1999) Assessing prepulse inhibition of startle in wild-type and knockout mice. Psychopharmacology(Berl) 147(1): 11–13
6. Ouagazzal AM, Jenck F, Moreau JL (2001) Drug-induced potentiation of prepulse inhibition of acoustic startle reflex in mice: a model for detecting antipsychotic activity? Psychopharmacology(Berl) 156(2–3):273–283
7. Arenas MC, Blanco-Gandía MC, Miñarro J, Manzanedo C (2020) Prepulse inhibition of the startle reflex as a predictor of vulnerability to develop. Front Behav Neurosci 13:296. https://doi.org/10.3389/fnbeh.2019.00296
8. Swerdlow NR, Light GA (2016) Animal models of deficient sensorimotor gating in schizophrenia: are they still relevant? Curr Top Behav Neurosci 28:305–325. https://doi.org/10.1007/7854_2015_5012
9. Swerdlow NR, Geyer MA (1998) Using an animal model of deficient sensorimotor gating to study the pathophysiology and new treatments of schizophrenia. Schizophr Bull 24: 285–302. https://doi.org/10.1093/oxfordjournals.schbul.a033326
10. Geyer MA, Krebs-Thomson K, Braff DL, Swerdlow NR (2001) Pharmacological studies of prepulse inhibition models of sensorimotor gating deficits in schizophrenia: a decade in review. Psychopharmacology 156:117–154. https://doi.org/10.1007/s002130100811
11. Joober R, Zarate JM, Rouleau GA, Skamene E, Boksa P (2002) Provisional mapping of quantitative trait loci modulating the acoustic startle response and prepulse inhibition of acoustic startle. Neuropsychopharmacology 27:765–781. https://doi.org/10.1016/S0893-133X(02)00333-0
12. Braff DL, Freedman R (2002) In: Davis KL, Charney D, Coyle JT, Nemeroff C (eds) Endophenotypes in studies of the genetics of schizophrenia. ACNP 5th Generation of Progress, Lippincott Williams and Wilkins, Philadelphia, PA, pp 703–716
13. Ibarra-Lecue I, Mollinedo-Gajate I, Meana JJ, Callado LF, Diez-Alarcia R, Urigüen L (2018) Chronic cannabis promotes pro-hallucinogenic signaling of 5-HT2A receptors through Akt/mTOR pathway. Neuropsychopharmacology 43(10):2028–2035. https://doi.org/10.1038/s41386-018-0076-y

Chapter 6

Automated Detection of Psychedelic-Induced Head-Twitch Response in Mice

Alaina M. Jaster and Javier González-Maeso

Abstract

Head-twitch response (HTR) allows for the detection and classification of behavior associated with serotonin 2A receptor (5-HT_{2A}R) activation upon psychedelic administration in rodent models. This activation and functional output can be utilized to provide insights into molecular mechanisms associated with psychosis and to identify signaling processes related to existing and novel antipsychotic and psychedelic compounds. Here we describe the use of a magnetic ear tag reporter coupled with automated quantification and biphasic detection to identify HTR in mice treated with the classical psychedelic 1-(2,5-dimethoxy-4-iodophenyl)-2-aminopropane (DOI).

Key words Head-twitch response (HTR), Behavioral model, Psychosis, Psychedelics, Hallucinogen, Schizophrenia, Serotonin 2A receptor (5-HT_{2A}R), Rodent models, Preclinical

1 Introduction

Head-twitch response (HTR) is a behavior noted in rodents in which the animal rapidly rotates the head side-to-side [1–3]. HTR is mainly studied in rats and mice, but a similar behavioral phenotype described as "wet-dog shakes" has also been noted in other mammals including rabbits, dogs, and cats [1, 4]. This behavior happens naturally in rodents following light ear touches, but the first ever recorded drug-elicited HTR was noted following intravenous administration of the psychedelic lysergic acid diethylamide (LSD), in which this behavior was rapidly occurring and sustained [2]. Several years following the discovery of this unique behavior, it was found that not only LSD but multiple psychedelics including psilocybin and mescaline produced HTR [5]. Since these original findings in the 1950s and 1960s, consistent reports have

Supplementary Information The online version contains supplementary material available at https://doi.org/10.1007/978-1-0716-3307-6_6.

Leyre Urigüen and Rebeca Diez-Alarcia (eds.), *Schizophrenia: Methods and Protocols*, Methods in Molecular Biology, vol. 2687, https://doi.org/10.1007/978-1-0716-3307-6_6,

demonstrated that this behavior increases in frequency following administration of serotonergic psychedelics, thus providing a physiological output associated with psychedelic action [1, 6–9] or a behavior model providing insights into molecular and neural circuit alterations relevant to neurodevelopmental and neuropsychiatric conditions [10–12].

Psychedelics are a class of compounds that, when taken by human subjects, cause changes in perception, cognition, and sensory processing [13]. The commonly reported subjective effects of psychedelics include visual hallucinations, euphoria, loss of sense of self, and some spiritual experiences [13]. Psychedelics, also known as classical hallucinogens, such as LSD, mescaline, and psilocybin, act primarily through serotonergic (5-HT) receptor activation—particularly via serotonin 2A receptor (5-HT_{2A}R)—to cause their subjective effects. In the early to mid-twentieth century, these compounds were noted to produce effects similar to symptoms reported by schizophrenia patients and were used by clinical researches to understand the mechanisms underlying psychosis in patients with this and other psychiatric conditions [14, 15]. More recent findings have raised concerns about the translational validity of psychedelics as a preclinical model of psychosis, but this family of psychoactive compounds did provide key insights into the pathology of schizophrenia mainly through the involvement of 5-HT [10, 16, 17]. Additionally, classical psychedelics as a preclinical model of psychosis also served as a tool in the search for new antipsychotics [11, 12, 17, 18].

The well-described and unrefuted mechanism behind psychedelic-induced HTR is thought to be via activation of the 5-HT_{2A}R in the frontal cortex [1, 7–9, 19–23]. Following the initial discovery of the LSD-induced HTR, the same researchers elicited HTR with the serotonin precursor, 5-hydroxytryptophan (5-HTP), thus proposing the involvement of 5-HT in the behavior [3]. It was later discovered that the 5-HT_2 receptors and, specifically, 5-HT_{2A}Rs were responsible for psychedelic-induced HTR in rodents [19, 20, 24]. Our group and others have corroborated the findings that the psychedelic-elicited HTR can be blocked using 5-HT_{2A}R antagonists such as ketanserin and volinanserin (M100907) [25, 26]. HTR is undetected in mice with genetic deletion of *5-$HT_{2A}R$* and the behavior is able to be rescued using Cre-mediated expression of *5-$HT_{2A}R$* in frontal cortex pyramidal neurons [27–29]. Analogous findings have been demonstrated in human subjects where ketanserin was found to prevent the subjective effects of psilocybin [11] and LSD [30].

We have demonstrated on multiple occasions the involvement of the 5-HT_{2A}R in psychedelic-induced HTR using both antagonists and knockout animals [9, 22, 23, 29]. Other groups have reported similar findings testing classical psychedelics as well as analogues and novel compounds [31–35]. The classical psychedelic and 5-HT_{2A}R agonist 1-(2,5-dimethoxy-4-iodophenyl)-2-amino-propane (DOI) has been reported to produce the most robust HTR when compared to other classical psychedelics such as LSD, psilocybin, and mescaline. Psilocybin and mescaline have counts significantly lower than LSD or DOI, and other classical psychedelics like N,N-dimethyltryptamine (DMT) produce unreliably low HTR counts [1, 4, 22]. Interestingly, structurally related non-hallucinogenic 5-HT_{2A}R agonists such as ergotamine and lisuride were not found to elicit the HTR [29]. It is important to note that some compounds that are not serotonergic also can produce HTR, like phencyclidine, and some cannabinoid receptor antagonists [36, 37]. The use of HTR has provided insights into the molecular target and mechanisms underlying psychedelic action and also antipsychotic drugs. Many antipsychotic medications are antagonists at both the dopamine D_2 receptors and 5-HT_{2A}Rs. Due to their effects at the 5-HT_{2A}R, some antipsychotics effectively block the induction of HTR induced by psychedelics [22, 38–41]. The consistency of findings across research groups solidifies that HTR is easily quantifiable, especially when automated, and has a low level of within-subject and between-subject variability [4].

With the rise of interest in understanding the targets and mechanisms behind the potential therapeutic effects of psychedelics in patients with psychiatric disorders such as depression and post-traumatic stress disorder (PTSD), the HTR behavior has become popularized as a functional tool to characterize psychedelic compounds. As discussed above, it is well established the involvement of 5-HT_{2A}R in the behavioral and cognitive effects of psychedelics as it relates to HTR and some subjective effects in humans, but it is not clearly established as to whether the effects of psychedelics on neuroplasticity processes, which may underlie the positive therapeutic effects of psilocybin and other psychedelic compounds, are 5-HT_{2A}R-dependent. There is considerable evidence from our lab [29, 42] and others [43, 44] suggesting that the psychedelics' effects on plasticity and certain behavioral outcomes associated with anxiety and depression are 5-HT_{2A}R-dependent, but other groups have suggested of 5-HT_{2A}R-independent mechanism for psilocybin [35, 45]. Interestingly, we and others have shown that structurally different psychedelics have different molecular targets when it comes to certain behaviors. For example, the effects of DOI on depression of intracranial self-stimulation (ICSS) in rats were found to be 5-HT_{2A}R-dependent, while effects of LSD and psilocybin were not [46]. Further, another group also reported that effects of DOI on compulsive-like behavior in mice were 5--HT_{2A}R-dependent, while effects of psilocybin were not [47]. The further characterization of these classical psychedelics, as well as the investigation of novel therapeutics, will rely heavily on a variety of preclinical models, including HTR.

As described elsewhere, HTR can be measured in a variety of ways: (i) video assessment and blinded scoring [4, 33], (ii) automated magnetometer detection [1, 22, 23], and (iii) automated video assessment [48]. Here, we describe the methodology for the DOI-induced HTR using an automated magnetometer detection via an ear tag reporter system that has been previously reported [1, 22, 23]. Using this method, we are able to record and transform HTR datasets from medium-throughput screening assays within approximately 24 h with consistent results across experiments.

2 Materials

2.1 Ear Tagging

1. Small neodymium magnets (N50, 3 mm diameter x 1 mm height, 50 mg).
2. Cyanoacrylate super glue.
3. Aluminum ear tags for rodents (Las Pias Ear Tag, Stoelting Co.)
4. Tag loading device.
5. Isoflurane delivery system.

2.2 Behavioral Apparatus and Detection System (Fig. 1)

1. 12.5 cm diameter plastic container surrounded by a coil (~500 turns 30 AWG enameled wire)
2. Pyle PP444 phono amplifier.
3. NI USB-6001 (National Instruments) data acquisition system.
4. MATLAB (Mathworks, R2018a version) driver with the corresponding National Instruments support package for further processing.

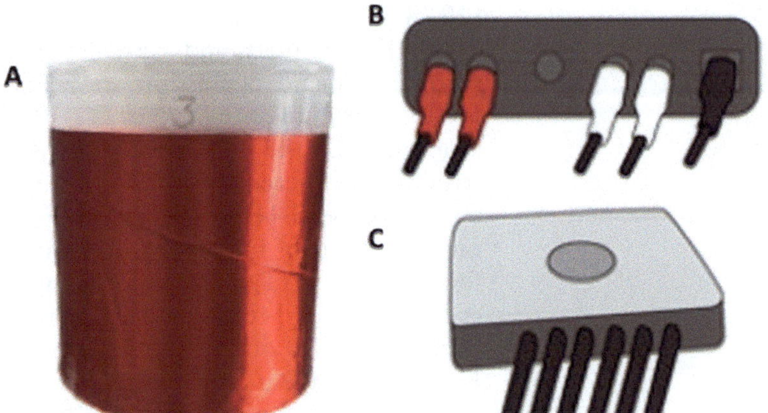

Fig. 1 Instrumentation. (**a**) Plastic container with coil, (**b**) schematic of amplifier, and (**c**) schematic of data acquisition system

5. MATLAB scripts and files (all found in supplementary materials).
   (a) Device driver—data acquisition.
   (b) Detector loader—data processing.
   (c) Detector script—data processing.
   (d) Correction manager—data processing.
   (e) HTR_T3_50.mat—pre-trained convolutional neural network.

2.3 Chemicals and Reagents

1. Isoflurane (2%).
2. 0.9% saline
3. DOI (DOI hydrochloride) diluted in saline to appropriate dose/volume (example: 1 mg/kg at 5 ul/g).
4. Diluted Rocal Spray.

3 Methods

Figure 2 shows a schematic outlining the workflow of the HTR assay.

3.1 Ear Tag Installation (Fig. 3)

1. Glue magnets onto ear tags N polarity facing down and allow 24 h to dry.
2. Check all magnets to make sure polarities are correct before installation, to optimize signal and prevent tags from sticking together.
3. Anesthetize mice under 2% isoflurane at a constant rate (*see* **Notes 1** and **2**).

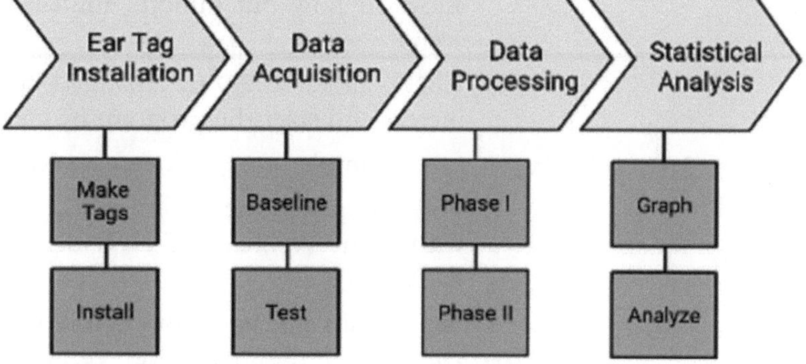

Fig. 2 Workflow diagram. Schematic outlining the workflow of the HTR assay. It is important to note that the data acquisition and processing are automated and following ear tag acclimation period, the rest of the workflow can be completed within 24 h for a single experiment

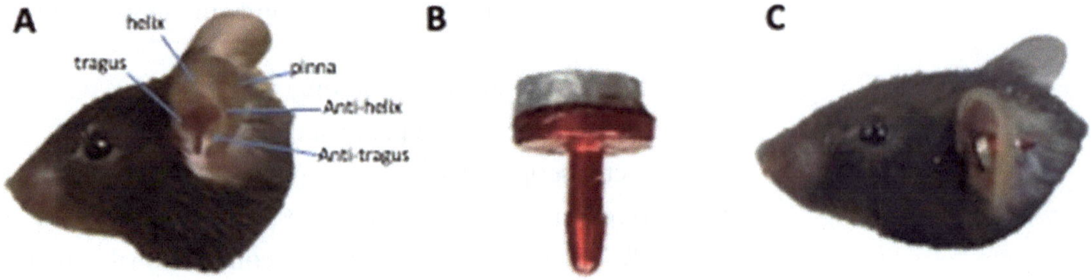

Fig. 3 Ear tag installation. (**a**) Anatomy of mouse ear, (**b**) ear tag, (**c**) mouse with ear tag

4. Insert the ear tag bilaterally through the pinna antihelix of both ears. The ear tags should be placed so that the magnet would rest in the interior of the antihelix, lying on top of the anti-tragus, leaving the ear canal unobstructed. The ear tags should be far enough apart that they do not stick together causing animal distress.
5. Return mice to home cage and allow at least 1 week to recover before conducting any experiments.

3.2 Behavioral Experiment and Data Acquisition

1. Set up the system by plugging the coils into the amplifier, which should be connected to the data acquisition system.
2. Open MatLab and the device driver file (see supplemental), and edit parameters for specific experiment.
3. Check the coils for magnetometer detection using a small magnet placed in each coil while program is running for 10 s. The wavelength of each coil should be present following that test.
4. Baseline 30-min recording sessions are always recommended before testing drug-induced HTR.
5. Place mice in corresponding coils and cover with a filtered lid to allow air flow but inhibits animals from climbing out of the arena.
6. Check all variables are changed corresponding to test parameters and start the program by clicking "run" on MatLab.
   (a) Example: *date_inititals_animalID_baseline.mat.*
7. Following baseline acclimation, inject mice intraperitoneally with saline (vehicle) or DOI using a 1 ml syringe and place back into coil.
8. Wait 5 min after injection before running the test.
9. Change the variables for the corresponding test and time to test DOI-induced HTR for 90 min.
   (a) Example: *date_inititals_animalID_treatment.mat.*
   (b) 90 min is recommended to assess peak effect of DOI, as well as other psychedelics.

10. When finished testing, remove mice and put back into home cages (*see* **Note 3**).
11. Spray and wipe down coils in between cohorts of mice. When finished with all experiments, spray and wipe out any feces or urine and gently rinse the inside with soap and water.
12. Recordings are processed during the behavioral experiment using a MATLAB (Mathworks, R2018a version) driver with the corresponding National Instruments support package for further processing. Experiments can be concurrently videotaped for visual identification of mouse behavior if needed. We have previously reported videotaping at 60 FPS and 1080p resolution. Detailed data acquisition has been reported previously [22, 23].

3.3 Data Processing

The necessary codes can be found in supplemental materials. For data processing, the detector loader and correction managers will be utilized. The HTR_T3_50 file must be present in the folder with the detector loader and correction manager, as it is a pre-trained convolutional neural network (*see* **Notes 4** and **5**).

1. Open the detector loader and enter in the file names from the data acquisition step. These files will then be processed using the medium–/high-throughput screening method.
2. Recorded signal is first band-passed through a finite impulse response (FIR) filter (70–110 Hz). The absolute values of the filtered signal were doubled, had the baseline subtracted, and were smoothed by applying a moving average. Maxima identification in the resulting peaks was used to screen for potential HTR.
3. Phase 1 identifies potential HTR via the following criteria:
   - (a) Have a magnitude exceeding a threshold value of 0.02 |V|.
   - (b) Have a ratio of prominence (maximum peak value relative to baseline) vs. peak height (maximum peak value relative to 0) below 0.95.
   - (c) Have a width (measured at half peak height) >20 and <150 ms.
   - (d) Is separated by a minimum of 200 ms from any other peak.
4. Phase 2 is a spectral density analysis, which events from Phase 1 undergo spectral density analysis that uses the following criteria to confirm HTR:
   - (a) Maximum in the 70–110 hz band of the spectrum exceeding a power threshold value of 0.005 V2/Hz.
   - (b) Cumulative sum of all power values in the above range in excess of 0.05 V2/Hz.

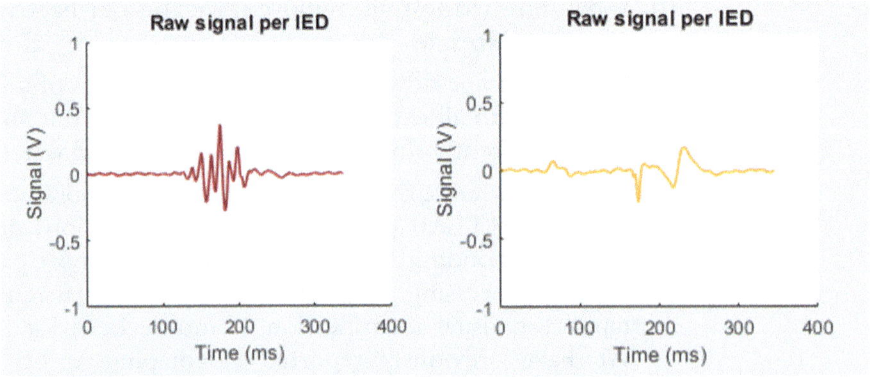

Fig. 4 Corresponding wavelengths to different behaviors identified using data processing. The wavelength shown in red is an example of a small psychedelic-induced HTR, which would be identified in Phase 1. The yellow wavelength is an example of random movement detected that would not be detected in as a potential HTR

   (c) Frequency of corresponding to the absolute maximum within the operative frequency range of the spectrum <35 Hz.

   (d) Sum of zero-crossings of the derivative of the spectrum density <40 Hz.

5. The script then produces folders of wavelengths to be visually inspected and sorted as false positives or false negatives (Fig. 4):

   (a) Any wavelength that resembles HTR upon visual inspection should be sorted into the corresponding true positives folder or those not resembling HTR should be moved into the true negatives folder.

6. Following visual inspection, the corresponding spreadsheet will be corrected using the correction manager and sorted based on the above parameters.

3.4 Basic Data Analyses

HTR data is typically analyzed via time course and total HTR using values interpolated from the data processing. Following processing, the script provides an XLS file with raw data and data from each processing step. The final processed data will be found in the fifth sheet of the file to be entered in statistical analysis software (i.e., GraphPad Prism).

1. For time course analysis, data should be entered in an xy format with the time points fractioned into time bins.

   (a) If using 15-min bins, baseline time points can be referred to as −15 and 0, or 0 and 15 (*see* Fig. 5). Baseline time points are typically no different between groups and are not included in data analysis.

   (b) Two-way ANOVA is used to assess main effect of DOI, time, and interaction.

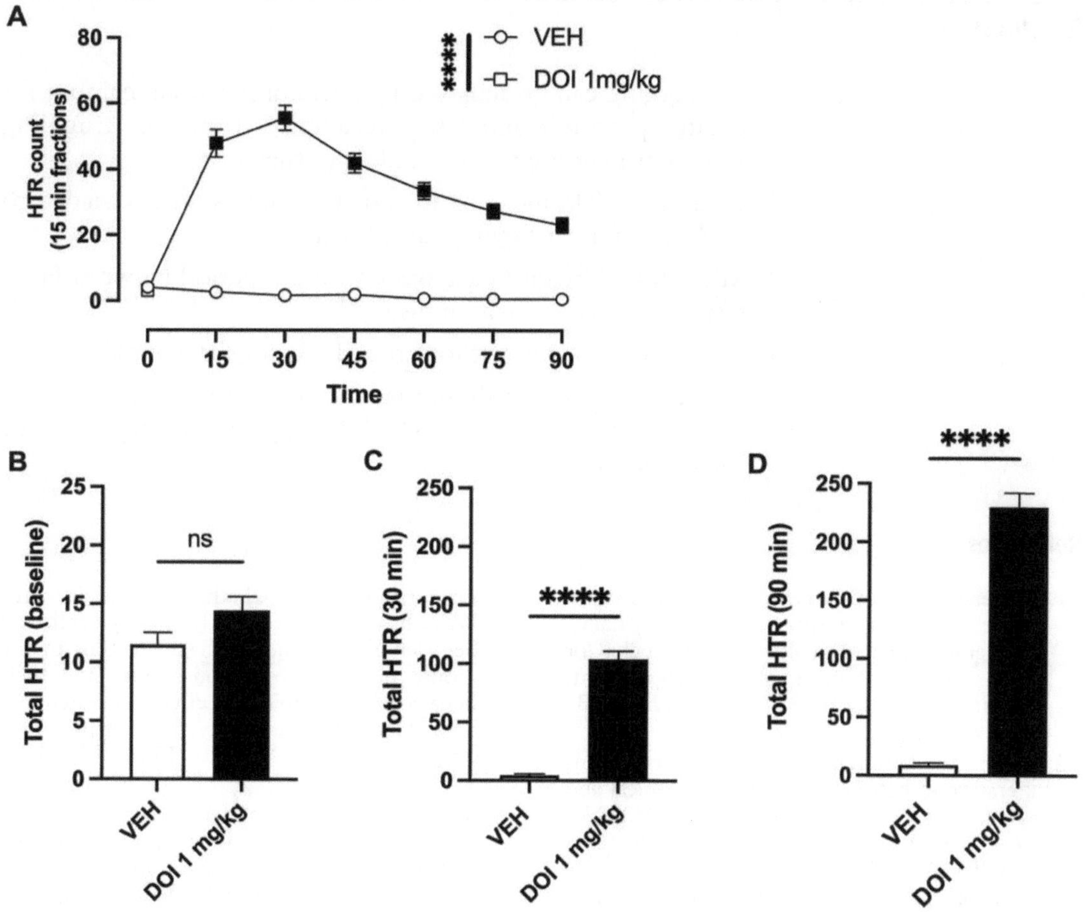

Fig. 5 DOI-induced HTR graphs and statistical analyses. (**a**) Time course of DOI-induced HTR over 90 min. (**b**) Total HTR at baseline, (**c**) 30 min and (**d**) 90 min. Time course (**a**) was analyzed using a two-way ANOVA to reveal effect of time (F[2.532, 55.70] = 46.54; $p < 0.0001$), DOI (F [1, 21] = 300.3; $p < 0.0001$) and interaction (F[6, 132] = 50.54; $p < 0.0001$). Bonferroni post hoc test revealed significant difference between DOI and saline (VEH) at filled in time points. Total HTR (**b–d**) was analyzed via two-tailed unpaired Student's *t*-test. ****$p < 0.0001$; ns, not significant. Data reported as mean ± SEM

2. For total HTR analysis, data can be entered in a column or grouped format depending on the number of variables. In the case of DOI vs. vehicle, a column format is used (Fig. 5).
   (a) If analyzing total HTR at 30-min peak, sum the total HTR in the appropriate time bins to obtain a single value per animal and enter in the column. This method is the same for assessing other times as well (90 min or baseline).
   (b) If just assessing a single dose of DOI vs. vehicle, Student's two-tailed unpaired *t*-test is used.
   (c) If dose response of DOI is being assessed, a one-way ANOVA would be utilized.

4 Notes

1. Ear tagging can be achieved by restraining the animal, but for greater precision and less interference from the struggling animal, we choose to use a mild anesthesia.
2. Ketamine/Xylazine mixture can be used as an alternative to isoflurane for ear tagging anesthesia.
3. Mice should be given a 1-week washout period between HTR tests when drugs are administered.
4. Scripts will not run without the HTR_T3_50.mat file.
5. All necessary scripts should remain in the same folder and be added to the path in MATLAB program or they will not run/process.

References

1. Halberstadt AL, Geyer MA. Characterization of the head-twitch response induced by hallucinogens in mice Detection of the behavior based on the dynamics of head movement. https://doi.org/10.1007/s00213-013-3006-z
2. Keller DL, Umbreit WW (1956) Permanent alteration of behavior in mice by chemical and psychological means. Science 124(3225): 723–724. https://doi.org/10.1126/SCIENCE.124.3225.723
3. Corne SJ, Pickering RW, Warner BT (1963) A method for assessing the effects of drugs on the central actions of 5-hydroxytryptamine. Br J Pharmacol Chemother 20(1):106–120. https://doi.org/10.1111/J.1476-5381.1963.TB01302.X
4. Canal CE, Morgan D (2012) Head-twitch response in rodents induced by the hallucinogen 2,5-dimethoxy-4-iodoamphetamine: a comprehensive history, a re-evaluation of mechanisms, and its utility as a model. Drug Test Anal 4(7–8):556–576. https://doi.org/10.1002/DTA.1333
5. Corne SJ, Pickering RW (1967) A possible correlation between drug-induced hallucinations in man and a behavioural response in mice. Psychopharmacologia 11(1):65–78. https://doi.org/10.1007/BF00401509
6. Halberstadt AL, Geyer MA (2014) Effects of the hallucinogen 2,5-dimethoxy-4-iodophenethylamine (2C-I) and superpotent N-benzyl derivatives on the head twitch response. Neuropharmacology 77:200–207. https://doi.org/10.1016/J.NEUROPHARM.2013.08.025
7. Halberstadt AL, Chatha M, Klein AK, Wallach J, Brandt SD (2020) Correlation between the potency of hallucinogens in the mouse head-twitch response assay and their behavioral and subjective effects in other species. Neuropharmacology:167. https://doi.org/10.1016/J.NEUROPHARM.2019.107933
8. de La Fuente RM, Shah UH, Nassehi N et al (2021) Psychedelic-like properties of quipazine and its structural analogues in mice. ACS Chem Neurosci 12(5):831–844. https://doi.org/10.1021/ACSCHEMNEURO.0C00291
9. Jaster AM, Elder H, Marsh SA et al. Effects of the 5-HT 2A receptor antagonist volinanserin on head-twitch response and intracranial self-stimulation depression induced by different structural classes of psychedelics in rodents. Psychopharmacology 1:3. https://doi.org/10.1007/s00213-022-06092-x
10. González-Maeso J, Sealfon SC (2009) Psychedelics and schizophrenia. Trends Neurosci 32(4):225–232. https://doi.org/10.1016/J.TINS.2008.12.005
11. Vollenweider Franz X, Vollenweider-Scherpenhuyzen Margreet FI, Andreas B, Helen V, Daniel H (1998) Psilocybin induces schizophrenia like psychosis in humans via serotonin-2 agonist action. Neuro Report 19: 2897–3902
12. de Gregorio D, Comai S, Posa L, Gobbi G (2016) D-Lysergic Acid Diethylamide (LSD) as a model of psychosis: mechanism of action and pharmacology. Int J Mol Sci 17(11). https://doi.org/10.3390/ijms17111953

13. Nichols DE (2016) Psychedelics. Pharmacol Rev 68(2):264–355. https://doi.org/10.1124/PR.115.011478
14. Rinkel M, De Shon HJ, Hyde RW, Solomon HC (2006) Experimental schizophrenia-like symptoms. https://doiorg/101176/ajp1088572 108(8):572–578. https://doi.org/10.1176/AJP.108.8.572
15. Halberstadt AL, Geyer MA (2013) Serotonergic hallucinogens as translational models relevant to schizophrenia. The international journal of neuropsychopharmacology/official scientific journal of the Collegium Internationale Neuropsychopharmacologicum (CINP) 16(10):2165. https://doi.org/10.1017/S1461145713000722
16. Muguruza C, Moreno JL, Umali A, Callado LF, Meana JJ, González-Maeso J (2013) Dysregulated 5-HT2A receptor binding in postmortem frontal cortex of schizophrenic subjects. Eur Neuropsychopharmacol 23(8):852. https://doi.org/10.1016/J.EURONEURO.2012.10.006
17. González-Maeso J, Ang RL, Yuen T et al (2008) Identification of a serotonin/glutamate receptor complex implicated in psychosis. Nature 452(7183):93–97. https://doi.org/10.1038/NATURE06612
18. Moreno JL, Miranda-Azpiazu P, García-Bea A et al (2016) Allosteric signaling through an mGlu2 and 5-HT2A heteromeric receptor complex and its potential contribution to schizophrenia. Sci Signal 9(410). https://doi.org/10.1126/SCISIGNAL.AAB0467
19. Glennon RA, Titeler M, McKenney JD (1984) Evidence for 5-HT2 involvement in the mechanism of action of hallucinogenic agents. Life Sci 35(25):2505–2511. https://doi.org/10.1016/0024-3205(84)90436-3
20. Darmani NA, Martin BR, Pandey U, Glennon RA (1990) Do functional relationships exist between 5-HT1A and 5-HT2 receptors? Pharmacol Biochem Behav 36(4):901–906. https://doi.org/10.1016/0091-3057(90)90098-3
21. Halberstadt AL, Geyer MA (2018) Effect of hallucinogens on unconditioned behavior. Curr Top Behav Neurosci 36:159–199. https://doi.org/10.1007/7854_2016_466
22. de la Fuente RM, Shin JM, Vohra HZ et al (2019) Fully automated head-twitch detection system for the study of 5-HT2A receptor pharmacology in vivo. Sci Rep 9(1). https://doi.org/10.1038/s41598-019-49913-4
23. de la Fuente RM, Vohra HZ, González-Maeso J (2020) Automated quantification of head-twitch response in mice via ear tag reporter coupled with biphasic detection. J Neurosci Methods 334:108595. https://doi.org/10.1016/J.JNEUMETH.2020.108595
24. Arnt J, Hyttel J (1989) Facilitation of 8-OHDPAT-induced forepaw treading of rats by the 5-HT2 agonist DOI. Eur J Pharmacol 161(1):45–51. https://doi.org/10.1016/0014-2999(89)90178-7
25. Offord SJ, Wong DF, Nyberg S (1999) The role of positron emission tomography in the drug development of M100907, a putative antipsychotic with a novel mechanism of action. J Clin Pharmacol 39(S1). https://doi.org/10.1002/J.1552-4604.1999.TB05933.X
26. Talvik-Lotfi M, Nyberg S, Nordström AL et al (2000) High 5HT(2A) receptor occupancy in M100907-treated schizophrenic patients. Psychopharmacology 148(4):400–403. https://doi.org/10.1007/s002130050069
27. López-Giménez JF, González-Maeso J, López-Giménez JF, González-Maeso J (2017) Hallucinogens and serotonin 5-HT 2A receptor-mediated signaling pathways. Curr Topics Behav Neurosci 36:45–73. https://doi.org/10.1007/7854_2017_478
28. González-Maeso J, Yuen T, Ebersole BJ et al (2003) Transcriptome fingerprints distinguish hallucinogenic and nonhallucinogenic 5-hydroxytryptamine 2A receptor agonist effects in mouse somatosensory cortex. J Neurosci 23(26):8836–8843. https://doi.org/10.1523/JNEUROSCI.23-26-08836.2003
29. González-Maeso J, Weisstaub NV, Zhou M et al (2007) Hallucinogens recruit specific cortical 5-HT2A receptor-mediated signaling pathways to affect behavior. Neuron 53(3):439–452. https://doi.org/10.1016/J.NEURON.2007.01.008
30. Preller KH, Razi A, Zeidman P, Stämpfli P, Friston KJ, Vollenweider FX (2019) Effective connectivity changes in LSD-induced altered states of consciousness in humans. Proc Natl Acad Sci U S A 116(7):2743–2748. https://doi.org/10.1073/PNAS.1815129116/-/DCSUPPLEMENTAL
31. Halberstadt AL, Chatha M, Chapman SJ, Brandt SD (2019) Comparison of the behavioral effects of mescaline analogs using the head twitch response in mice. J Psychopharmacol 33(3):406. https://doi.org/10.1177/0269881119826610
32. Halberstadt AL, Geyer MA (2014) Effects of the hallucinogen 2,5-dimethoxy-4-iodophenethylamine (2C-I) and superpotent N-benzyl derivatives on the head twitch response. Neuropharmacology 0:200–207. https://doi.org/10.1016/J.NEUROPHARM.2013.08.025

33. Buchborn T, Lyons T, Knöpfel T (2018) Tolerance and tachyphylaxis to head twitches induced by the 5-HT2A agonist 25CN-NBOH in mice. Front Pharmacol 9(FEB):17. https://doi.org/10.3389/FPHAR.2018.00017/BIBTEX
34. Halberstadt AL (2020) Automated detection of the head-twitch response using wavelet scalograms and a deep convolutional neural network. Sci Rep 10(1). https://doi.org/10.1038/S41598-020-65264-X
35. Hesselgrave N, Troppoli TA, Wulff AB, Cole AB, Thompson SM. Harnessing psilocybin: antidepressant-like behavioral and synaptic actions of psilocybin are independent of 5-HT2R activation in mice. https://doi.org/10.1073/pnas.2022489118/-/DCSupplemental
36. Darmani NA, Pandya DK (2000) Involvement of other neurotransmitters in behaviors induced by the cannabinoid CB1 receptor antagonist SR 141716A in naive mice. J Neural Transm (Vienna) 107(8–9):931–945. https://doi.org/10.1007/S007020070043
37. Nabeshima T, Ishikawa K, Yamaguchi K, Furukawa H, Kameyama T (1987) Phencyclidine-induced head-twitch response in rats treated chronically with methysergide. Eur J Pharmacol 133(3):319–328. https://doi.org/10.1016/0014-2999(87)90028-8
38. Bozymski KM, Lowe DK, Pasternak KM, Gatesman TL, Crouse EL (2017) Pimavanserin: a novel antipsychotic for Parkinson's disease psychosis. Ann Pharmacother 51(6):479–487. https://doi.org/10.1177/1060028017693029
39. Meltzer HY, Mills R, Revell S et al (2010) Pimavanserin, a serotonin 2A receptor inverse agonist, for the treatment of Parkinson's disease psychosis. Neuropsychopharmacology 35(4):881–892. https://doi.org/10.1038/npp.2009.176
40. Wang RY, Liang X (1998) M100907 and clozapine, but not haloperidol or raclopride, prevent phencyclidine-induced blockade of NMDA responses in pyramidal neurons of the rat medial prefrontal cortical slice. Neuropsychopharmacology 19(1):74–85. https://doi.org/10.1016/S0893-133X(98)00003-7
41. Park WK, Jeong D, Cho H et al (2005) KKHA-761, a potent D3 receptor antagonist with high 5-HT 1A receptor affinity, exhibits antipsychotic properties in animal models of schizophrenia. Pharmacol Biochem Behav 82(2):361–372. https://doi.org/10.1016/j.pbb.2005.09.006
42. de la Fuente RM, Zhu B, Guevara CA et al (2021) Prolonged epigenomic and synaptic plasticity alterations following single exposure to a psychedelic in mice. Cell Rep 37(3):109836. https://doi.org/10.1016/J.CELREP.2021.109836
43. Ly C, Greb AC, Cameron LP et al (2018) Psychedelics promote structural and functional neural plasticity. Cell Rep 23(11):3170. https://doi.org/10.1016/J.CELREP.2018.05.022
44. Cameron LP, Benson CJ, Defelice BC, Fiehn O, Olson DE. Chronic, intermittent microdoses of the psychedelic N,N-Dimethyltryptamine (DMT) produce positive effects on mood and anxiety in rodents. Published online 2019. https://doi.org/10.1021/acschemneuro.8b00692
45. Shao LX, Liao C, Gregg I et al (2021) Psilocybin induces rapid and persistent growth of dendritic spines in frontal cortex in vivo. Neuron 109(16):2535–2544.e4. https://doi.org/10.1016/J.NEURON.2021.06.008
46. Jaster AM, Elder H, Marsh SA, de la Fuente RM, Negus SS, González-Maeso J (2022) Effects of the 5-HT2A receptor antagonist volinanserin on head-twitch response and intracranial self-stimulation depression induced by different structural classes of psychedelics in rodents. Psychopharmacology 1:1–13. https://doi.org/10.1007/S00213-022-06092-X/FIGURES/5
47. Odland AU, Kristensen JL, Andreasen JT (2021) Investigating the role of 5-HT2A and 5-HT2C receptor activation in the effects of psilocybin, DOI, and citalopram on marble burying in mice. Behav Brain Res 401:113093. https://doi.org/10.1016/J.BBR.2020.113093
48. Glatfelter GC, Chojnacki MR, McGriff SA, Wang T, Baumann MH. Automated Computer Software Assessment of 5-Hydroxytryptamine 2A Receptor-Mediated Head Twitch Responses from Video Recordings of Mice. ACS Pharmacology & Translational Science. Published online April 8, 2022: acsptsci.1c00237. https://doi.org/10.1021/ACSPTSCI.1C00237

Chapter 7

Gastrointestinal and Microbiome Profiling in Rodent Models of Neurological and Psychiatric Disorders

Geraldine Kong, Carolina Gubert, and Anthony J. Hannan

Abstract

Mice and other rodent models have been widely used to understand the role of the gut microbiome in various neurological and psychiatric disorders. Here we describe a protocol to characterize the structural and functional phenotype of the rodent gut and to examine the gut microbiota composition through V4 16S rRNA gene sequencing and microbiome profiling. This protocol will have utility for those investigating the gut, and associated microbiota, in a wide range of different rodent models of human disorders.

Key words Gut microbiome, V4 16S rRNA sequencing, Gut permeability, Gastrointestinal transit time, Brain disorders

1 Introduction

Changes in gut microbiome composition are associated with a wide spectrum of neurological disorders. Growing knowledge on mediators and modulators has advanced our understanding of the cross-talk between the microbiome, the gastrointestinal tract, and the brain. While diet is one of the most critical factors in governing the gut microbiome composition, the host environment, and the gastrointestinal tract in particular, it also plays a role in shaping the microbial community. The gut mucosa lining (consisting of a dense mucus layer, epithelial cells, lymphoid and the enteric nervous system) acts as a protective barrier for the integrity of the intestinal tract, as well as regulating gut motility and gut microenvironment, which is known to impact gut microbiome composition [1, 2]. On the other hand, commensal microorganisms produce metabolites that impact the structural integrity and gut motility through the same aforementioned systems [3]. Thus, it is not surprising that gastrointestinal symptoms are common comorbidities in

Authors Geraldine Kong and Carolina Gubert have equally contributed to this chapter.

Leyre Urigüen and Rebeca Diez-Alarcia (eds.), *Schizophrenia: Methods and Protocols*, Methods in Molecular Biology, vol. 2687, https://doi.org/10.1007/978-1-0716-3307-6_7,

neurological and psychiatric disorders [4]. Understanding both the gut microbiome composition and gut physiology is crucial in uncovering new therapeutic approaches for disease modulation. Here we describe methods to characterize some of the functional phenotypes of the rodent gastrointestinal tract, as well as profiling the gut microbiome using V4 16S rRNA sequencing, adapted from Earth Microbiome Project's protocol.

2 Materials

2.1 Carmine Red Dye Solution

1. Mix carmine red dye (Sigma-Aldrich, C1022) as a 6% (w/v) dilution in 0.5% methylcellulose (Sigma-Aldrich, M0512).
   1.1. Prepare 0.5% methylcellulose under 80 °C. Heat one-third of the final water volume before adding the methylcellulose powder.
   1.2. Add the remaining volume of water (2/3) once the particles are dissolved.
   1.3. Cool the solution down to 0–5 °C for 20–40 min.
   1.4. Store at 4 °C until use (stable up to 1 year).
   2.1. Add carmine to the 0.5% methylcellulose solution heating until the carmine dissolves.
   2.2. Store at 4 °C until use (stable up to 1 year).
   3. Autoclave and filter the final carmine red dye solution before administration.

2.2 4 kDa Fluorescein Isothiocyanate (FITC)-Dextran

1. Dissolve the 4 kDa fluorescein isothiocyanate (FITC)-dextran (Sigma-Aldrich, FD4) in 1x PBS, at a final concentration of 100 mg/mL, wrapping it in aluminum foil.
2. Filter and sterilize the solution.
3. Tubes wrapped with aluminum foil should be stored at −20 °C until use (stable for 6 months) or 4 °C (stable for 3–4 weeks).

2.3 Fecal Genomic DNA Extraction

1. QiAamp PowerFecal Pro Dna kit (QIAGEN or equivalent).

2.4 V4 16S rRNA Library Amplification and Cleanup

1. PCR primers 515F (with Golar barcode) and 806R.
2. 2X Platinum Hot Start PCR Master Mix (ThermoFisher).
3. Nuclease-free water.
4. Thermocycler.
5. Quant-iT dsDNA assay kit, high sensitivity (ThermoFisher).
6. SPRIselect beads (Beckman Coulter).
7. 100% ethanol.
8. Microfuge tubes.
9. Magnetic rack.

2.5 16S rRNA Library Denaturation and Sequencing

1. 0.2 M sodium hydroxide.
2. PhiX control.
3. Forward, reverse, and index sequencing primer.

3 Methods

3.1 Structural and Functional Phenotyping of the Rodent Gut

3.1.1 Fecal Output and Water Content

Fecal water content is considered a measure of gastrointestinal water absorption and a reflection of constipation. Fecal output is considered a marker of gut function and motility.

1. Acclimatize the mice to the procedure/behavior room for 30 min prior to the test.
2. Single house the animals in clean, sterilized, and empty boxes for 60 min.
3. During the 60-min period: every 15 min, count, record the numbers, and collect the excreted pellets, putting them into sealed identified Eppendorf tubes (*see* **Note 1**).
4. Record the weight of the tubes containing the total number of pellets collected during the 60-min period. The final number of pellets excreted is taken as the fecal output.
5. Following the 60-min testing period, return all the animals to their home cage.
6. Place the tubes (with the lids opened) in a dry heating block overnight at 65 °C kept in a fume hood (to avoid the unpleasant smell) (*see* **Note 2**).
7. Record the weight of the tubes containing the dried pellets.
8. The percentage difference between the initial total feces weight and the dry weight is taken as the fecal water content, calculated as follows:

$$[(\text{total wet weight (g)} - \text{total dry weight (g)}) / \text{total stool wet weight (g)}] \times 100]$$

3.1.2 Gastrointestinal Transit Time

Gut transit time is considered a marker of gut function and motility.

1. Gavage non-fasted mice with 150 μL of the carmine red 6% (w/v) in 0.5% methylcellulose solution and record the time of administration. Single house them for 9 h in clean boxes containing facial tissue and ad libitum water and food (*see* **Notes 3** and **4**).
2. Check the boxes often (at least every 15 min after 1.5 h post gavaging) and record the time that the first red pellet appears (*see* **Notes 5** and **6**).
3. The time taken from the gavage to the first appearance of a red fecal pellet is recorded as the gastrointestinal transit time.
4. After 9 h, place all the mice back to their home cages (*see* **Note 7**).

3.1.3 FITC-Dextran Intestinal Permeability

FITC-dextran intestinal permeability reflects intestinal epithelial barrier permeability and can be associated to "leaky gut."

1. Fast mice for 4 h at the beginning of their light cycle (or overnight) (*see* **Notes 8** and **9**).
2. Prepare the solution of 4 kDa fluorescein isothiocyanate (FITC)-dextran (*see* **Note 10**).
3. Gavage the mice with 150 μL of the 100 mg/mL FITC-dextran solution recording the time of injection (*see* **Notes 11** and **12**).
4. After 4 h of FITC-dextran administration, collect the blood via cardiac puncture, followed by the euthanasia method approved by your institution's animal ethics committee.
5. Immediately transfer the collected blood to an EDTA collection tube.
6. Centrifuge the EDTA tubes at 1,000× g for 10 min.
7. Collect plasma and transfer to identified Eppendorf tubes (*see* **Notes 13** and **14**).
8. Sample processing for fluorescence quantification:
   - We recommend performing a pilot beforehand to identify the limit of detection for FITC and appropriate dilution factor that applies to your conditions.
   - Perform a FITC-dextran serially dilution in PBS a standard curve (0, 125, 250, 500, 1,000, 2,000, 4,000, 8,000 ng/mL).
   - Use plasma from mice that did not receive FITC-dextran administration to determine the background of the samples.
   - Add 100 μL of standard curve, diluted samples, and background control to an opaque black 96-well plate in duplicate.
9. Quantify the fluorescence at an excitation wavelength of 485 nm and an emission wavelength of 528 nm using a plate reader (*see* **Note 15**).

3.1.4 Macroscopic Measures

Gut macroscopy measures are indicators of general gut health status. In addition, the ratio between the colon weight and the colon length can be considered an indirect assessment of gut inflammation [5]. In addition, previous studies exploring gut macroscopy established that germ-free mice have enlarged cecums, while preclinical IBS studies have found an enlarged cecum and shorter colon length [6, 7]. Antibiotic-treated mice also show an enlarged cecum [8], suggesting that cecum size is associated the gut microbiota population.

1. Euthanize the mice by a method approved by your institution's ethics committee.

2. Remove the gastrointestinal tract from the cecum to the rectum, placing gently on a non-absorbent surface (*see* **Notes 16** and **17**).
3. Place a ruler close to the gastrointestinal tract and record the length of the cecum and colon (*see* **Note 18**).
4. Using a razor or a scissor, separate the cecum from the colon.
5. Weigh the full cecum (*see* **Note 19**).
6. Clean and flush the colon tissue before weighing it (*see* **Notes 20** and **21**).

3.2 Profiling the Gut Microbiome

3.2.1 Fecal Sample Collection for 16S rRNA Sequencing

1. Place each mouse individually into sterile cages until 3–4 fecal pellets are excreted (*see* **Note 22**).
2. Collect the fresh fecal pellets aseptically using needles and place into identified 1.5 mL Eppendorf tubes (*see* **Note 23**).
3. Immediately snap frozen with dry ice or liquid nitrogen.
4. Store at −80 °C until the analysis (*see* **Note 24**).

3.2.2 Fecal Genomic DNA Extraction

1. Extract genomic DNA material from one fecal pellet, roughly one rice grain size, per sample using QIAamp PowerFecal Pro DNA kit (QIAGEN) according to the manufacturer's instructions (*see* **Notes 25–28**).

3.2.3 16S rRNA V4 Amplicon Library Preparation

1. Dilute the PCR primers to a working concentration of 10 μM. These include the 515F primer, containing the unique 12 nt Golay barcode as per the Earth Microbiome Project's protocol, and the 806R primer, which amplify the V4 hypervariable region of the 16S rRNA gene [9].
2. Thaw Platinum Hot Start PCR Master Mix 2x (Thermo Fisher) on ice. Mix by inverting each tube, briefly spin down, and keep the tube on ice.
3. Prepare 25 μL PCR reaction on ice consisting of:
   - (a) 2x Platinum Hot Start PCR Master Mix – 10 μL
   - (b) 515F forward primer, barcoded (10 μM) – 0.5 μL
   - (c) 806R reverse primer (10 μM) – 0.5 μL
   - (d) Template DNA – 1 μL
   - (e) Nuclease-free water – 13 μL
4. Use the following thermocycler conditions for PCA amplification:
   - (a) Denaturation: 94 °C – 3 min
   - (b) Melting: 94 °C – 45 s
   - (c) Annealing: 50 °C – 60 s
   - (d) Extension: {72 °C – 90 s} × 35 cycles
   - (e) Final extension: 72 °C – 10 min
   - (f) Incubation: 10 °C – ∞

5. Quantify library concentration using Quant-iT dsDNA high sensitivity assay kit (ThermoFisher) according to the manufacturer's protocol (*see* **Note 29**).
6. Pool equal concentrations of amplicon library from each sample into a single tube. Vortex to mix well and briefly centrifuge.
7. Cleanup the PCR reaction by adding 0.5 volume of SPRI select beads (Beckman Coulter) to the pooled libraries. Mix by inverting the tube ten times. Incubate at room temperature for 5 min (*see* **Note 30**).
8. Briefly spin down the sample. Keep the tube on the magnetic rack and wait until the liquid is clear (~3 min). Carefully remove the supernatant.
9. With the tube containing the pooled libraries on the magnetic rack, wash the beads three times as follows:
   (a) Add 500 μL of freshly prepared 80% ethanol to the tube.
   (b) Incubate for 30 s.
   (c) Carefully remove and discard the supernatant without disturbing the beads.
10. Use a 20 μL pipette to remove and discard residual ethanol. Air dry the pellet for 2 min (*see* **Note 31**).
11. Remove the tube from the magnetic rack and resuspend pellet in 200 μL of PCR-grade water. Invert ten times and incubate at room temperature for 2 min.
12. Keep the tube on a magnetic rack until the liquid is clear.
13. Transfer 200 μL supernatant to a clean tube (*see* **Note 32**).

3.2.4 Preparing DNA Amplicon Libraries for Sequencing

1. Quantify the cleaned library pool using Quant-iT dsDNA high sensitivity assay kit (ThermoFisher) according to the manufacturer's protocol (*see* **Note 33**).
2. Calculate the DNA concentration (nM) of the pool according to the following formula:

$$\text{Concentration (nM)} = \frac{\left(\text{Sample Concentration}\left(\frac{\text{ng}}{\mu\text{L}}\right) \times 10^6\right)}{660 \times 390}$$

3. Dilute the pool to a concentration of 2 nM with 10 mM Tris pH 8.5. Vortex and briefly centrifuge the tube.
4. To prepare 6.5 pM of amplicon libraries:
   (a) Denature the libraries by combining 2 nM pooled library (10 μL) with 0.2 N sodium hydroxide (10 μL) in a clean microfuge tube. Mix by pipetting gently and incubate at room temperature for 5 min.
   (b) Quench the reaction with 980 μL of chilled HT1 solution provided in the MiSeq kit. Vortex the sample, now at 20 pM, and rest on ice.

   (c) Dilute the sample to 6.5 pM by adding 162.5 μL of 20 pM of denatured libraries and 337.5 μL of HT1 in a new microfuge tube. Rest on ice until use.

5. To prepare 6.5 pM of PhiX control:
   (a) Dilute 10 nM of PhiX to 2 nM by adding 2 μL of PhiX and 8 μL of 10 mM Tris pH 8.5 in a clean microfuge tube.
   (b) Denature the PhiX by adding 10 μL of 0.2 N freshly prepared sodium hydroxide. Mix by pipetting gently and incubate at room temperature for 5 min.
   (c) Quench by adding 980 μL of HT1. Vortex the PhiX, now 20 pM, to mix and rest on ice.
   (d) Dilute the PhiX to 6.5 pM by adding 162.5 μL of 20 pM of PhiX and 337.5 μL of HT1 in a new microfuge tube. Rest on ice until use.
6. To prepare the final loading amplicon library with 15% PhiX spike in (*see* **Note 34**), combine 150 μL of 6.5 pM PhiX control and 850 μL of 6.5 pM amplicon library. Rest on ice until use.

3.2.5 Sequencing of Amplicon Libraries on MiSeq Platform

1. Thaw a MiSeq reagent cartridge in a water bath at room temperature for about 1 h prior use.
2. Remove the MiSeq cartridge from the water bath and dry the sides with paper towel. Invert the cartridge a few times to mix the reagent and tap the cartridge down to ensure the reagents have settled at the bottom.
3. Load the MiSeq cartridge as follows using Illumina sequencing primers:
   (a) Pierce well 12 and load 4 μL of 100 μM Read 1 sequencing primer.
   (b) Pierce well 13 and load 4 μL of 100 μM index primer.
   (c) Pierce well 14 and load 4 μL of 100 μM Read 2 sequencing primer.
   (d) Pierce well 17 and load 600 μL of the final loading amplicon library containing the PhiX.
4. Clean the MiSeq flow cell by:
   (a) Gently rinsing it with nuclease-free water until thoroughly rinsed of excess salts.
   (b) Drying the flow cell with a lint-free lens cleaning tissue while using care around the black flow cell port gasket.
   (c) Cleaning the flow cell glass with an alcohol wipe to wipe off excess streaks, fingerprints, or tissue fibers while avoiding the flow cell gasket.
   (d) Drying excess alcohol with a lint-free lens cleaning tissue.
5. Place the flow cell on the stage of the MiSeq platform.

6. Mix the PR2 bottle by inverting, remove the lid, and load the bottle in the reagent compartment door.
7. Empty the contents of the waste bottle into the appropriate waste container.
8. Lower the sipper handle and ensure that the sippers lower into the PR2 and waste bottle.
9. Load the MiSeq reagent cartridge and start the sequencing run using a sample sheet (.csv) file containing the run parameters, ensuring that the file is named as the specific barcode on the MiSeq cartridge.

3.2.6 Data Analysis Using QIIME2

1. Install QIIME2 (Quantitative Insights into Microbial Ecology 2) following the steps on their website (https://docs.qiime2.org/) [10].
2. Activate QIIME2.
3. The following assumes that the sequencing was performed paired-end 150 bp multiplexed data. Sequence data should be in FASTQ format with R1 reads renamed as forward.fastq.gz, R2 reads renamed as reverse.fastq.gz, and the index reads renamed as barcodes.fastq.gz. Ensure that all sequencing files are in the same directory named "emp-paired-end-sequences":
   (a) forward.fastq.gz,
   (b) reverse.fastq.gz,
   (c) barcodes.fastq.gz.
4. Import the raw sequences into QIIME2 environment.

```
qiime tools import \
 --type EMPPairedEndSequences \
 --input-path emp-paired-end-sequences \
 --output-path emp-paired-end-sequences.qza
```

5. Demultiplex the sequences to know which barcode sequence is associated with each sample. The following code assumes that the barcodes are in the tab-delimited file "barcode_file.tsv" under the column "barcode_sequence." The demultiplexed sequences are stored in "demux.qza," and the details about the barcode error correction are stored in "demux-details.qza."

```
qiime demux emp-paired \
 --i-seqs emp-paired-end-sequences.qza \
 --m-barcodes-file barcode_file.tsv \
 --m-barcodes-column barcode_sequence \
 --o-per-sample-sequences demux.qza \
 --o-error-correction-details demux-details.qza \
 --p-rev-comp-mapping-barcodes \
 --p-rev-comp-barcodes
```

6. After demultiplexing, it is useful to generate a summary of the demultiplexing results to know how many sequences were obtained per sample and also to get a summary of the distribution of sequence qualities at each position in the sequence data.

```
qiime demux summarize \
 --i-data demux.qza \
 --o-visualization demux.qzv
```

To visualize the summary of the demultiplexing results:

```
qiime tools view demux.qzv
```

7. Filter the sequences based on quality scores. The filtered reads and the filtering statistics are stored in "demux-filtered.qza" and "demux-filter-stats.qza," respectively.

```
qiime quality-filter q-score \
 --i-demux demux.qza \
 --o-filtered-sequences demux-filtered.qza \
 --o-filter-stats demux-filter-stats.qza
```

8. Generate a count matrix through denoising by Deblur. This requires the "p-trim-length" parameter that truncates the sequences at the specified position and should be adjusted according to the quality plot in "demux.qzv" obtained above. The output "table.qza" contains the count matrix, whereas the "rep-seqs.qza" contains the actual sequences of the resolved amplicon sequence variants (ASVs).

```
qiime deblur denoise-16S \
 --i-demultiplexed-seqs demux-filtered.qza \
 --p-trim-length 150 \
 --o-representative-sequences rep-seqs.qza \
 --o-table table.qza \
 --p-sample-stats \
 --o-stats deblur-stats.qza
```

The two commands below are used to visualize the summary statistics generated above:

```
qiime metadata tabulate \
--m-input-file demux-filter-stats.qza \
 --o-visualization demux-filter-stats.qzv

qiime deblur visualize-stats \
--i-deblur-stats deblur-stats.qza \
 --o-visualization deblur-stats.qzv
```

9. The following command generates a visual summary of the denoised data and assumes that the sample metadata file is named "sample-metadata.tsv," containing samples as rows and metadata in columns. The "feature-table summarize"

command gives the information on how many sequences are associated with each sample and each feature, while the "feature-table tabulate-seqs" command provides a mapping of the feature IDs to their exact sequence. This is useful to know the number of sequences for rarefaction that will be used for the phylogenetic analysis.

```
qiime feature-table summarize \
 --i-table table.qza \
 --o-visualization table.qzv \
 --m-sample-metadata-file sample-metadata.tsv
 qiime feature-table tabulate-seqs \
 --i-data rep-seqs.qza \
 --o-visualization rep-seqs.qzv
```

To visualize .qzv files:

```
qiime tools view table.qzv
qiime tools view rep-seqs.qzv
```

10. Construct a phylogenetic tree from the representative sequences. This is required in order to compute phylogenetic diversity measures such as Faith's phylogenetic diversity or unweighted and weighted UniFrac. QIIME2 uses MAFFT for multiple sequence alignments, followed by the masking stage that removes alignment positions that do not contain enough conservation to provide meaningful information. The tree is then built using FastTree and is midpoint rooted, resulting in "rooted-tree.qza," which is the input for the phylogenetic diversity measures.

```
qiime phylogeny align-to-tree-mafft-fasttree \
 --i-sequences rep-seqs.qza \
 --o-alignment aligned-rep-seqs.qza \
  --o-masked-alignment masked-aligned-rep-seqs.qza
\
 --o-tree unrooted-tree.qza \
 --o-rooted-tree rooted-tree.qza
```

11. Generate an alpha rarefaction curve that provides information on how extensive a community has been sampled. The following command computes multiple alpha diversity metrics at various sampling depths, up to the value provided for the parameter "--p-max-depth."

```
qiime diversity alpha-rarefaction \
 --i-table table.qza \
  --i-phylogeny rooted-tree.qza \
  --p-max-depth 10000 \
  --m-metadata-file sample-metadata.tsv \
  --o-visualization alpha-rarefaction.qzv
```

Visualize the alpha-rarefaction.qzv file to determine the sampling depth for the core-metrics-phylogenetic test in the following step:

```
qiime tools view alpha-rarefaction.qzv
```

12. Perform alpha and beta diversity analysis with the following command lines. This will generate multiple alpha diversity metrics including Shannon's diversity index, observed features, and Faith's phylogenetic diversity and evenness, as well as beta diversity metrics including Jaccard, Bray-Curtis, unweighted UniFrac, and weighted UniFrac distance. These can be found in the folder specified in the "--output-dir" parameter. The "p-sampling-depth" parameter can be adjusted according to the "table.qzv" and the "alpha-rarefaction.qzv" generated above. If the total of counts for any sample(s) is smaller than this value, those samples will be dropped from the diversity analysis. It is recommended to choose a value as high as possible to retain more sequences per sample while excluding as few samples as possible.

```
qiime diversity core-metrics-phylogenetic \
 --i-phylogeny rooted-tree.qza \
 --i-table table.qza \
 --p-sampling-depth 1500 \
 --m-metadata-file sample-metadata.tsv \
 --output-dir phylo_res
```

Principal coordinates (PCoA) plots generated from the above commands through Emperor can be used to explore microbial community composition in the context of sample metadata. To view the PCoA plot:

```
qiime tools view phylo_res/bray_curtis_emperor.
qzv
qiime tools view phylo_res/jaccard_emperor.qzv
qiime tools view phylo_res/unweighted_unifrac_em-
peror.qzv
qiime tools view phylo_res/weighted_unifrac_em-
peror.qzv
```

The following command line is an example code for testing the associations between categorial metadata columns and Faith's phylogenetic diversity through the Kruskal-Wallis test:

```
qiime diversity alpha-group-significance \
  --i-alpha-diversity phylo_res/faith_pd_vector.
qza \
```

```
--m-metadata-file sample-metadata.tsv \
  --o-visualization phylo_res/faith-pd-group-sig-
nificance.qzv
```

Similarly, the following command line is an example of how to investigate sample composition in relation to sample metadata using PERMANOVA. It tests whether the unweighted UniFrac distances between samples with the same diet are more similar to each other than they are to samples fed with a different diet. The "--m-metadata-column" parameter can be adjusted to the categorical-metadata-of-interest. To perform pairwise tests that allow the determination of which specific pairs of groups differ to another, add in the "--p-pairwise" parameter.

```
qiime diversity beta-group-significance \
  --i-distance-matrix phylo_res/unweighted_uni-
frac_distance_matrix.qza \
  --m-metadata-file sample-metadata.tsv \
  --m-metadata-column diet \
  --o-visualization phylo_res/unweighted-unifrac-
diet-significance.qzv \
  --p-pairwise
```

13. Assign taxonomy to representative sequences by aligning them to Greengenes reference database or any other desired database matching with the region of the sequenced amplicons.

```
qiime feature-classifier classify-sklearn \
 --i-classifier gg-13-8-99-515-806-nb-classifier.
qza \
 --i-reads rep-seqs.qza \
 --o-classification taxonomy.qza

qiime metadata tabulate \
 --m-input-file taxonomy.qza \
 --o-visualization taxonomy.qzv
```

Generate a taxonomic barplot to visualize the taxonomic composition of the samples interactively. It is optional to provide a sample metadata sheet for this.

```
qiime taxa barplot \
 --i-table table.qza \
 --i-taxonomy taxonomy.qza \
 --m-metadata-file sample-metadata.tsv
 --o-visualization tax-bar-plots.qzv
```

To visualize the interactive taxonomic barplots:

```
qiime tools view tax-bar-plots.qzv
```

4 Notes

1. If the fecal pellets are not collected every 15 min and kept in a tube with a lid, it can compromise the analysis, since it may lead to a difference between the first and the last excreted pellets—the fecal water starts evaporating as soon as they are out. The pellets that make direct contact with urine should be counted but not collected, since this will interfere in the final water content measurement.
2. Alternatively keep at 95 °C for >3 h.
3. Use plastic needles to minimize the stress associated to the gavaging procedure.
4. It is possible to fast the mice overnight before the test, but the impact and need for this additional, and stressful, procedure should be considered. We recommend controlling for the food intake of your mice prior to the test.
5. Smear any potential red pellet with a white tissue to facilitate the identification.
6. While the first red pellet can appear anytime after the gavaging, all the mice should be single housed for the 9 h of the experiment to avoid different times of isolation, making sure all the mice are exposed to the same conditions.
7. The mice that do not excrete any red pellet during this period should be added to the analysis as >540 min.
8. Fasting the mice overnight is more convenient considering the length of the experiment and that the endpoint is euthanasia. However, the inherent impact of this procedure on the mice should be considered, and importantly, all the conditions should be the same for all mice included in the study.
9. If fasting the mice overnight, record their weight prior to culling in case of further macroscopic measures being analyzed, since they might be corrected by body weight.
10. FITC-dextran is sensitive to light and temperature (we suggest preparing a fresh solution, but the solution can be kept at 4 °C for up to 4 weeks).
11. Use plastic needles to minimize the stress associated with the gavaging procedure.
12. We recommend gavaging each mouse with intervals of 15–30 min considering the time needed for the blood collection, euthanasia, and potential tissue collection.
13. Keep the blood and plasma samples covered by foil, and at 4 °C, until FITC quantification.

14. We suggest quantifying the samples on the same day as the blood collection, but we observed that they can be kept at 4 °C for up to 3 days before quantification without any signal interference—however please note that all the plates should be read under the same conditions (straight from the refrigerator to the reader or after spending the same number of hours on the bench).
15. We use PHERAstar FSX, Millipore.
16. Make sure to perform this step as quickly as possible without stretching the tissue, since this can interfere with the measurements.
17. We recommend to keep the tissue wet with 0.1% of nicardipine (an L-type Ca2+ channel blocker that inhibits smooth muscle contraction) in PBS 1x.
18. A photo can be taken with proper identification and the lengths determined later.
19. Cecum content can be collected for further 16S RNA sequencing analysis.
20. A syringe with a plastic needle can be used with PBS 1x for the flushing step.
21. The cecum and colon length and weight can be normalized to body weight (g).
22. We recommend having as many boxes as possible per experimental group in order to minimize the batch effect of the box due to the rodents coprophagic nature that lead to sharing gut microbiome among cagemates.
23. The pellets that make direct contact with urine should not be collected.
24. We recommend avoiding long-term storage and freeze-thaw cycles.
25. Samples can be homogenized using Precellys 24 Tissue Homogenizer (Bertin Instruments) at 5,000 rpm × 30 s.
26. Elute the DNA in 85 μL of Solution C6 (QiAmp PowerFecal Pro DNA kit), incubating for 1 min at room temperature.
27. There may be less DNA extracted from fecal pellets of antibiotic-treated mice and, if this is the case, adjust the elution volume as necessary.
28. Store the DNA frozen, as the elution buffer does not contain EDTA, and avoid multiple freeze-thaw cycles.
29. Alternatively, run 1 μL of the PCR product on an Agilent Bioanalyzer DNA 1000 chip or equivalent to verify the library size profile.

30. For example, add 100 μL of beads to 200 μL of the pooled libraries.
31. Do not overdry the beads to the point of cracking. They should be wet but not shiny.
32. The A260/280 ratio of the final, cleaned pool should be between 1.8 and 2.0 for best results.
33. The pool should be between 0.5 ng/μL and 5 ng/μL.
34. Illumina recommends using ≥25% PhiX control spike-in if using an older version of the MiSeq software.

References

1. McVey Neufeld KA, Perez-Burgos A, Mao YK et al (2015) The gut microbiome restores intrinsic and extrinsic nerve function in germ-free mice accompanied by changes in calbindin. Neurogastroenterol Motil 27(5):627–636. https://doi.org/10.1111/nmo.12534. Epub 2015 Mar 2
2. Herath M, Hosie S, Bornstein JC et al (2022) The role of the gastrointestinal mucus system in intestinal homeostasis: implications for neurological disorders. Front Cell Infect Microbiol 10:248. https://doi.org/10.3389/fcimb.2020.00248. PMID: 32547962; PMCID: PMC7270209
3. Vicentini FA, Keenan CM, Wallace LE et al (2021) Intestinal microbiota shapes gut physiology and regulates enteric neurons and glia. Microbiome 9(1):210. https://doi.org/10.1186/s40168-021-01165-z. PMID: 34702353; PMCID: PMC8549243
4. Camilleri M (2021) Gastrointestinal motility disorders in neurologic disease. J Clin Invest 131(4):e143771. https://doi.org/10.1172/JCI143771. PMID: 33586685; PMCID: PMC7880310
5. Carvajal AE, Vázquez-Carretero MD, García-Miranda P et al (2017) Reelin expression is up-regulated in mice colon in response to acute colitis and provides resistance against colitis. Biochim Biophys Acta Mol basis Dis 1863(2):462–473. https://doi.org/10.1016/j.bbadis.2016.11.028. Epub 2016 Nov 30
6. Al-Asmakh M, Zadjali F (2015) Use of germ-free animal models in microbiota-related research. J Microbiol Biotechnol 25(10):1583–1588. https://doi.org/10.4014/jmb.1501.01039
7. Ohgo H, Imaeda H, Yamaoka M et al (2016) Irritable bowel syndrome evaluation using computed tomography colonography. World J Gastroenterol 22(42):9394–9399. https://doi.org/10.3748/wjg.v22.i42.9394. PMID: 27895427; PMCID: PMC5107703
8. Ge X, Ding C, Zhao W et al (2017) Antibiotics-induced depletion of mice microbiota induces changes in host serotonin biosynthesis and intestinal motility. J Transl Med 15(1):13. https://doi.org/10.1186/s12967-016-1105-4. PMID: 28086815; PMCID: PMC5237163
9. Caporaso JG, Ackermann G, Apprill A et al (2018). EMP 16S Illumina amplicon protocol. See http://www.earthmicrobiome.org/protocols-and-standards/16s
10. Bolyen E, Rideout JR, Dillon MR et al (2019) Reproducible, interactive, scalable and extensible microbiome data science using QIIME 2. Nat Biotechnol 37(8):852–857. https://doi.org/10.1038/s41587-019-0209-9. Erratum in: Nat Biotechnol 2019;37(9):1091. PMID: 31341288; PMCID: PMC7015180

Chapter 8

Electroencephalography for the Study of the Auditory P300 Evoked Potential and Derived Measurements

Rosa M. Beño-Ruiz-de-la-Sierra, Inés Fernández-Linsenbarth, Alejandro Roig-Herrero, and Álvaro Díez-Revuelta

Abstract

Electroencephalography (EEG) is a widely used tool in neuropsychiatry research. The most used measurements in EEG are the amplitude and latency of the cortical electrophysiological activity in response to stimulus, known as evoked potentials. Besides potentials, time/frequency analysis is also used to obtain information on global fluctuations of the recordings, which evoked potentials do not provide. Time/frequency analysis results in different values known as derived measures. In this work, a brief introduction to evoked potentials and time/frequency analyses in schizophrenia is given, focusing on P300, noise power, and spectral entropy. Finally, a detailed description is given on how to obtain EEG recordings, evoked potentials, and derived measures.

Key words Electroencephalography, Evoked potentials, P300, Time-frequency, Spectral entropy, Noise power

1 Introduction

1.1 Electroencephalography

Electroencephalography (EEG) is a neuroimaging technique that collects information mainly from surface cortical activity using electrodes placed on the scalp. EEG can provide real-time data on the state of the central nervous system [1] and has had—and continues to have—a major impact on the study of brain functions in both healthy individuals and neuropsychiatric conditions. Compared to other neuroimaging techniques, EEG has the main advantages of offering high temporal resolution—in terms of milliseconds—it can be used noninvasively and does not require a large space or financial outlay to set up the required equipment.

1.2 Event-Related Potentials

Event-related potentials (ERPs) are neuroelectrical responses of the cortex evoked by a stimulus that has been repeatedly presented or emitted by the brain as it initiates its response [2, 3]. Thus, ERPs

Leyre Urigüen and Rebeca Diez-Alarcia (eds.), *Schizophrenia: Methods and Protocols*, Methods in Molecular Biology, vol. 2687, https://doi.org/10.1007/978-1-0716-3307-6_8,

provide information on the activity of the human brain when perceiving or making decisions. Compared to background EEG activity, ERPs are very low-voltage fluctuations (1–30 μV) that appear due to the summation of many action potentials (APs) related to sensory, motor, or cognitive events [3]. Each waveform can be analyzed in terms of amplitude (the extent of neural activity, usually in microvolts or μV), latency (the timing in milliseconds when the peak amplitude occurs), and scalp distribution (the pattern of voltage gradient over the scalp) [2, 4].

ERPs can be labeled as exogenous or endogenous depending on whether, respectively, they are influenced by the physical attributes of the repeated stimulus (occurring within 100 ms after each onset) or by the psychological effects of the stimulation [2, 3]. Their nomenclature is based on polarity (amplitude above or below basal activity) and peak latency [2, 3]. Thus, positive ERPs are labeled by the letter P, whereas negative potentials are represented by the letter N, followed by digits indicating the approximate milliseconds of potential onset after the stimulus presentation (i.e., P300 or N100).

1.3 P300

The P300 wave is a positive ERP whose latency can be located at about 300 ms after stimulus onset and is thought to reflect high-level cognitive processes (i.e., related to an attention and working memory task). The P300 mainly includes two distinct components: P3a is related to frontal attentional mechanisms (i.e., automatic orientation toward novel or salient stimuli), whereas P3b evokes tempo-parietal activity related to memory processing (i.e., maintenance toward and instructed target) [5].

The oddball paradigm has typically been used to elicit the P300 potential. In this paradigm, task-relevant infrequent targets appear amid task-irrelevant frequent stimuli, and subjects are instructed to respond only to the former. This paradigm has been studied in both healthy individuals and clinical populations. One of the strongest findings is that people with schizophrenia have altered P300 potential, mainly presented as a significantly lower amplitude and a later latency [6].

1.4 Time-Frequency Analysis

Time-frequency analysis comprises several methods and measurements that capture different aspects of EEG magnitude and phase relationships [7], reflecting the brain's parallel processing of information [8]. Evoked potential activity can be broken down into different frequency bands, by using the Fourier transform, and measured by its amount of energy (or power, in μV^2) and phase in relation to the stimulus presentation. The most widely used frequency bands are delta (1–4 Hz), theta (4–8 Hz), alpha (8–13 Hz), beta (13–30 Hz), and gamma (30–70 Hz) [9]. This allows analysis of the amount of neural activity that is related (evoked) or unrelated (induced) to the stimulus onset, and some of the latter can be considered as noise activity [10].

The "noise power" (NP) measure could be defined as activity independent of stimulus presentation during an evoked potential paradigm and has been previously studied in patients with schizophrenia during an oddball paradigm to elicit the P300 potential [11–14]. Compared to healthy controls, patients with schizophrenia have shown significantly increased gamma-band NP values in frontal midline, and this alteration correlated with higher negative symptoms [15] and cognitive deficits [11, 12, 16]. This finding can be interpreted as an excess of task-irrelevant, nonspecific activity in patients, which is in line with evidence of impaired neural synchronization in schizophrenia [17] and may reflect an imbalance of excitatory and inhibitory neuronal activity proposed as a key pathophysiological element in this disorder [18].

1.5 Entropy

Among the EEG-derived measures potentially relevant to the study of schizophrenia, "spectral entropy" (SE) is a useful parameter for quantifying the degree of disorder contained in a signal. Shannon [19] defined entropy as the average amount of information in a probability distribution. Later, this concept was extended to the EEG field in terms of power spectral density (PSD) [20]. Specifically, the SE calculation consists of an estimation of the flatness of the spectral content in the EEG signal [21, 22]. Low SE values indicate a spectrum with a narrow frequency range (i.e., a more regular signal), whereas high SE values imply a uniform spectrum with a broad spectral content (i.e., a more irregular signal). Figure 1 illustrates the behavior of the SE through a simulation using a synthetic signal.

The regularities of EEG activity in different frequency bands likely play an important role in the coordination of activity across cortical regions [23–25]. Thus, through the comparison of average signal variability across time between groups, SE allows an indirect assessment of the mechanisms underlying cognitive processing. In this line, previous studies have shown that SE modulation during an auditory oddball task (i.e., signal order during P300 evocation minus signal at baseline) is decreased in patients with schizophrenia compared to healthy controls in the broadband [9, 26, 27]. This result means that the EEG signal becomes more regular (SE levels are reduced) during task performance in healthy controls, but this process is deficient in people with schizophrenia. This result was unrelated to pharmacological treatment [28] and may reflect an altered adaptation or synchronization capacity of the neural assemblies underlying cognition in schizophrenia under task demands.

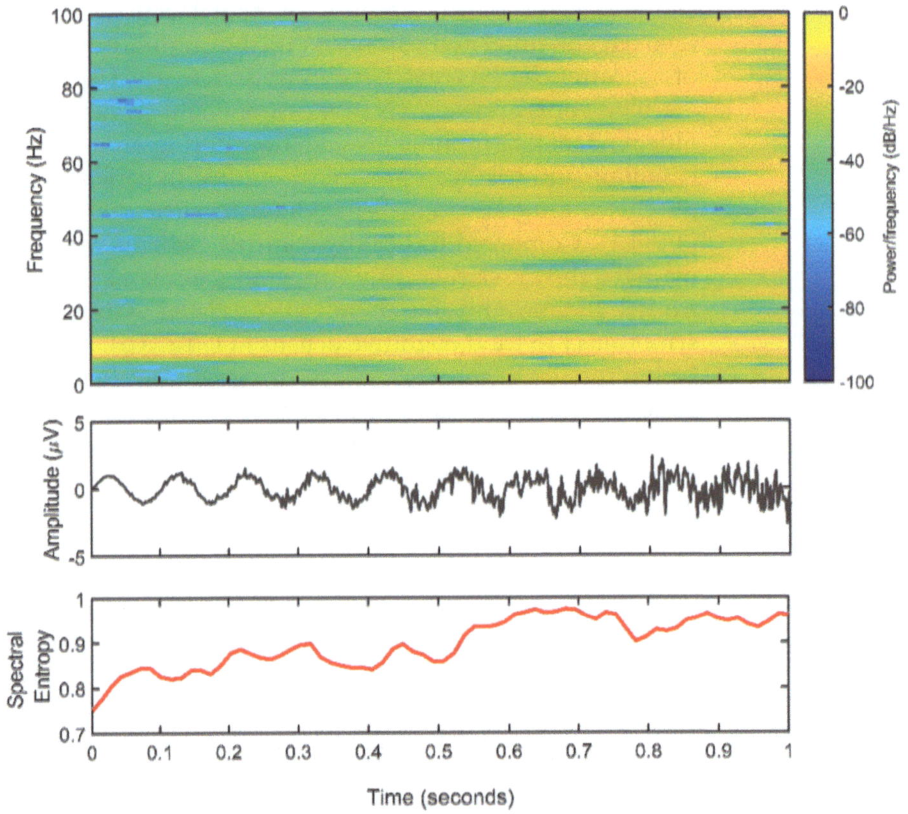

Fig. 1 Simulation of SE behaviour using a synthetic signal. As frequency bands add to the ongoing spectrum (upper part), signal complexity (middle) and spectral entropy (lower part) increase.

2 Materials

2.1 Laboratory Conditions

EEG recording should be conducted in a room that is isolated or minimizes electromagnetic or acoustic noise, as these can cause significant artifacts in the EEG signal. The room should be free of visual distractions for the participant (e.g., posters, pictures, windows, etc.). Sufficient space should be provided for research staff to circulate around the participant's position, as well as a table for the preparation of material. The equipment used in the laboratory of the SUCEDE (Brain Substrates of Psychosis) research group is described below.

2.2 Recording Equipment

1. A computer with monitor and Brain Recorder software (Brain Products, Gilching, Germany) installed (license key required).
2. EEG recording equipment, consisting of an amplifier and a power supply battery, including appropriate connectors.
3. Connector cable between recording computer and EEG amplifier with USB adapter.

4. A 64-electrode EEG recording cap with mounting according to the international 10/10 system (electrodes FP1, FP2, F7, F8, F3, F4, Fz, FC5, FC6, FC1, FC2, T7, T8, C3, Cz, C4, CP5, CP6, CP1, CP2, TP9, TP10, P7, P8, P3, P4, Pz, O1, O2, Oz, AF7, AF3, AF3, AF10, P7, P8, P3, P4, Pz, O1, O2, Oz, AF7, AF3, AFz, F1, F5, FT7, FC3, FCz, C1, C5, TP7, CP3, P1, P5, PO7, PO3, POz, PO4, PO8, P6, P2, CPz, CP4, TP8, C6, C2, FC4, FT8, F6, F2, AF4, AF8).
5. Two single electrodes for electrooculogram (EOG) recording (horizontal and vertical).
6. Consumables for cap and electrode preparation: conductive gel, syringes and blunt needles, electrode stickers, alcohol, cleaning paper, and cotton wool.
7. A comfortable chair or armchair with a backrest and cushions of different sizes for the participant's comfort during EEG registration.

2.3 Stimulation Equipment

1. A computer with monitor and Presentation software (Neuro-Behavioral Systems Inc., Albany CA) installed (license key required).
2. Connector cable between stimulation computer and EEG amplifier with USB adapter.
3. Two external loudspeakers for listening to auditory stimulation.

2.4 Material for Signal Processing

A computer with Matlab software (MathWorks Inc., MA, USA) installed (license required), as well as Matlab Signal Analysis (MathWorks Inc., MA, USA), EEGlab [29], and Fieldtrip [30] toolboxes for EEG signal processing.

3 Methods

3.1 Lab and Material Setup

1. Check that you have all the material (equipment, cables, and consumables).
2. Ensure that the battery of the amplifier and any other necessary devices (e.g., if using a laptop for recording and/or stimulation) is sufficiently charged (*see* **Note 1**).
3. Start the computers and the specific recording (Brain Recorder) and stimulation (Presentation) programs.
4. Insert both software USB licenses (or another way provided by your supplier).
5. Advise the participant in advance to come to the session with comfortable clothing, without hair accessories or make-up on the forehead and eyes, and clean hair free of conditioner, hairspray, or hair gel that could hinder the lowering of impedances in the electrodes.

3.2 EEG and EOG Preparation

1. Seat the participant in a comfortable chair (*see* **Note 2**) facing the two loudspeakers, which shall be positioned symmetrically and equidistant from the participant. The distance to the recording amplifier must be sufficient so that the connection cables to the EEG cap are sufficiently loose (*see* **Note 3**).
2. Select the most appropriate cap size for the participant's size and place it on the participant's head. This action will be done mainly with four hands. The most efficient way is to place the front part first, press against the forehead, and then extend the side and back areas of the cap without releasing the front area. Finally, extend the fastening strap under the chin to secure the cap, ensuring that it exerts sufficient pressure but no discomfort or tension on the participant (*see* **Note 4**).
3. Check that the cap is properly fitted and of the appropriate size. With the tape measure, measure the distance between inion and nasion (100%). Then check that the measurements between nasion and prefrontal electrodes (Fpz), between inion and occipital electrodes (Oz), and between preauricular point and temporal electrodes (T9 and T10) are each approximately 10% of the first overall measurement. The Cz electrode will be at the vertex point.
4. Connect the EEG cap to the recording amplifier and enable impedance display in the Brain Recorder program (Brain Products, Gilching, Germany) of the recording computer.
5. Open de conductive gel container, collect gel in the syringe, and then attach the blunt needle.
6. Select a first electrode, insert the needle into the hole, and inject some gel. Normally the ground and reference electrodes are prepared first (*see* **Note 5**).
7. In our experience, the most efficient procedure to lower the electrode impedance is to make 360-degree turns with the syringe while the needle contacts the scalp. These turns should be made at different angles of inclination. After several turns, inject a new quantity of gel. The amounts of gel injected each time should be small (equivalent to one or two drops) in order to prevent excessive spreading of gel on the scalp and to avoid possible coupling with another electrode. Continue the procedure until all impedances have been reduced to 5 micro-ohms or below (*see* **Note 6**).
8. For the preparation of the EOG electrodes, the upper and lateral external area of the participant's right eyebrow is cleaned with a cotton swab and alcohol. The EOG electrodes are then attached to these areas approximately 1 cm away from the eyebrow. Gel is introduced and the impedance is lowered similarly to the EEG electrodes.

3.3 Stimulating Presentation Paradigm

1. The EEG recording is obtained during the performance of a stimulation and cognitive task and following the three-stimulus oddball paradigm. This consists of the presentation of a target tone (500 Hz; probability 0.2), a distractor tone (1,000 Hz; probability 0.2), and a standard tone (2,000 Hz; probability 0.6). The participant must respond only to the target tone—by clicking the mouse—and avoid a response to the other two types of tone.
2. A shorter test trial should be carried out before the experimental one. In this manner, the participant learns to distinguish between the three tones. Feedback must be given to the participant during this test trial, and it must be repeated until the task is understood (i.e., effectively respond only to target tones).
3. The experimental trial consists of 600 binaural tones presented in random order with an inter-stimulus interval randomly fluctuating between 1.16 and 1.44 s. This paradigm extends for more than 13 min and the participant is told to keep eyes closed (*see* **Note** 7).
4. Stimulus presentation is carried out in the Presentation software (NeuroBehavioral Systems Inc., Albany CA) (on the stimulation computer), which must be connected and synchronized with the recording computer via the EEG amplifier (as described in Subheading 2.3). The following information is synchronized and automatically recorded: the times of stimulus presentation and participant response, as well as each type of stimulus presented.
5. For further analyses, only the response to correctly attended target tones is used (i.e., targets followed by participant's response). Therefore, subjects who do not respond correctly to at least 50% of the stimuli should not be included in further analyses.

3.4 EEG Recording

1. Once the impedances have been sufficiently reduced, activate the EEG display viewer in the Brain Recorder software. The recording will be observed according to the selected montage. The setup and channel montage must have been established beforehand. Our equipment uses the Cz reference and two additional channels for EOG.
2. On the stimulation computer, prepare the paradigm (test trial or final experiment) in the Presentation program (NeuroBehavioral Systems Inc., Albany CA). The stimulation routine must have been prepared in advance in this program.
3. Three sound files (corresponding to standard, distractor, and target tones) will be prepared and shown to the participant before the start for recognition.

4. Read the following instruction to the participant: "*Next, we are going to do a listening task. It is important that you concentrate on doing your best. You should do the task with your eyes closed and relaxed so as not to make any body or* facial *muscle movements that interfere with the recording. I am going to present you with 3 different sounds, pay attention: this is the high-pitched sound* [play sound 1], *this is the medium-pitched sound* [play sound 2] *and this is the low-pitched sound* [play sound 3]. *Now you will listen to these 3 sounds randomly. You should only attend to the lowest of the 3* [play sound 3 again] *and press the mouse button each time* [press the mouse button]. *Any questions?* [clarify any doubts]. *Let's start!*"
5. Simultaneously activate the stimulation paradigm and the EEG recording on both computers. Insert a predefined participant identifier code when prompted by the software (under which the information will be stored).
6. Stand away from the participant so that he/she is not aware of our presence (ideally in a Gesell chamber where we can observe the performance), and ensure silence during the test.
7. During the recording, be attentive to the participant's performance and give additional instructions if necessary (e.g., in case of inattention or muscle movements).
8. Once the paradigm is finished, carefully remove the cap from the participant and provide a convenient means of cleaning his/her head. Wash the cap and other materials appropriately according to the manufacturer's recommendations.

3.5 Signal Pre-processing

After recording, the EEG data must be exported from the Brain Recorder software (Brain Products, Gilching, Germany) (each subject will have a separate record and file under his identification code) to another computer where Matlab software (MathWorks Inc., MA, USA) is installed. Specific toolboxes such as EEGlab [29] or Fieldtrip [30] can be used for signal pre-processing.

The purpose of pre-processing is twofold. First, it can be used to reorganize the data without losing any information in order to facilitate its analysis. Second, it is used to partially remove the noise included in the recording. It is worth noticing that it is not possible to remove noise from the recording without removing some data at the same time. Therefore, a balance should be found for every experiment depending on how much data you have.

Pre-processing steps usually includes:

1. *Filtering:* Certain amplitudes (related to artifact) are directly removed from the signal by means of specific filters. We use a *bandpass filter* for the frequencies of interest (delta to gamma, i.e., from 1 to 70 Hz) [9]. A second *notch filter* corresponds to the alternating current (AC) oscillations to eliminate its effect (50 Hz in Europe, 60 Hz in America).

2. *Cleaning or artifact removal through independent component analysis (ICA):* ICA is a statistical technique that attempts to identify independent sources of variance in the data. This allows some artifacts to be isolated so that they are removed from the signal. The most common artifact components are those related to the eyes (blinking and saccadic movement) and muscle tension or movement (*see* **Note 8**).
3. *Epoching or segmentation:* The interest of evoked potentials lies in the data recorded just after the presentation of the stimulation (around 300 ms for the P300 potential). Therefore, recordings should be segmented into windows or epochs of a specific duration around the stimulus presentation. These windows typically span a pre-stimulus baseline and up to sufficient time after stimulus onset to ensure collection of the full evoked potential. Our protocol typically acquires a −300 ms to 700 ms window [6] (no further for avoidance of the next stimulus onset). These segmentations should be performed separately for distractor (P3a) and target tones (P3b). It is not performed for standard tones.

3.6 Evoked Potential Processing

There are multiple steps that must be performed to obtain the evoked potentials from the pre-processed signal. These steps must be done for each electrode separately:

1. *Baseline correction:* The goal of this step is to place all epochs on the same scale. To this end, the pre-stimulus signal (i.e., −300 to 0 ms [31]) is averaged and then subtracted from the rest of the epoch. Consequently, after the baseline correction, all epochs fluctuate around 0 μV.
2. *Epoch rejection:* Despite the cleaning done in the pre-processing, obvious artifacts can still be found during the calculation of the potential. These epochs often show large high-frequency amplitudes that usually correspond to subject movement during the EEG recording. These epochs with artifact can be removed by automatic methods (exceeding a threshold such as 70 μV) or manual methods (by visual inspection).
3. *Averaging trials per subject:* The evoked potential is only visible when removing all background (random) activity not related to stimulus onset. This is achieved by averaging the signal across trials. This step should be performed for the deviant and target tone independently (P3a and P3b, respectively). The higher the number of stimulation repetitions per condition, the clearer the P300 potential will show.
4. *Averaging across subjects (grand averages):* The resulting individual P3a and P3b potentials can be further averaged per group (e.g., healthy controls and patients with schizophrenia separately) to obtain the grand averages.

5. *Amplitude calculation:* It can be calculated in two alternative ways: [1] as the maximum amplitude (in μV) corresponding to the P300 peak in a predefined window around 300 ms after stimulus onset (e.g., 150–450 ms [31]) and [2] as the average of the amplitude in this window. The advantage of the second method is that it is less exposed to singular amplitude values that may distort the P300 magnitude; however, the disadvantage is that we will not be able to calculate the latency.
6. *Latency calculation:* It is defined as the time between the stimulus onset (deviant for P3a and target for P3b) and the peak or maximum amplitude value corresponding to the P300 potential.

3.7 Calculation of Derived Measures

Derived EEG measures require the implementation of their calculation in specific software such as Matlab (MathWorks Inc., MA, USA) with the help of toolboxes such as EEGlab [29] and Fieldtrip [30].

3.7.1 Noise Power (NP)

The total power (expressed in μV^2) is calculated by subjecting the segmented data to spectral analysis using the fast Fourier transform (FFT). The noise power can be calculated for the different bands by pre-filtering the signal for each frequency range (delta to gamma). We can then calculate the noise power following previous recommendations [14, 32], based on the signal-to-noise ratio (SNR), a measure of the quality of the EEG signal applied to each frequency band. SNR and total power values are computed for the desired time and frequency windows using the Brain Vision software (Brain Products, Gilching, Germany). The time window selected can be 300 to 400 ms for the P3a evoked potential (earlier potential) and 300 to 600 for the P3b evoked potential (later potential), or use a visual method for the selection of the window that best matches the resulting P300 potential [12, 16, 33, 34].

The average noise power (ANP) can be calculated for each individual participant, band, electrode, and task condition (standard, deviant, and target stimuli). Since in the EEG recording neither the signal nor the noise powers are exactly known, ANP must be estimated with statistical methods. The signal power corresponds to the mean-squared amplitude of a series of the averaged trials, so that it is estimated by calculating the mean power of the averaged evoked single trials [10, 35]. ANP can then be computed by subtracting this averaged signal power (ASP) from the averaged total power (ATP) of the single sweep, while SNR represents the quotient of the ASP divided by the ANP:

$$\mathrm{ANP} = \mathrm{ATP} - \mathrm{ASP}, \quad \mathrm{SNR} = \frac{\mathrm{ASP}}{\mathrm{ANP}}.$$

Then, ANP can be estimated by:

$$\mathrm{ANP} = \frac{\mathrm{ATP}}{\mathrm{SNR} + 1}.$$

3.7.2 Spectral Entropy (SE)

Another approach for characterizing the EEG signal is by using the power spectral density (PSD) function, which represents how the power is distributed in the frequency domain. The short-time Fourier transform (STFT), a sliding temporal window technique used to obtain the time evolution of the PSD [26, 36], can be used to assess the time-frequency maps of ERP signals:

$$\mathrm{PSD}_n^{(i)}(f) = \frac{\mathrm{PSD}^{(i)}(f)}{\sum_{f=1\mathrm{Hz}}^{70\mathrm{Hz}} \mathrm{PSD}^{(i)}(f)},$$

where $\mathrm{PSD_n}$ is the normalized PSD function, i is the number of time intervals, and 1–70 Hz is the frequency interval selected. $\mathrm{PSD_n}$ can be considered as a probability distribution in the band of interest [1 70] Hz. This representation provides a suitable tool for the application of several spectral parameters.

Next, to calculate the spectral entropy (SE) values, Shannon's definition of entropy is applied to the computed $\mathrm{PSD_n}$:

$$SE^{(i)} = -\frac{1}{\log(L)} \cdot \sum_{f=1\mathrm{Hz}}^{70\mathrm{Hz}} \mathrm{PSD}_n^{(i)}(f) \cdot \log\left[\mathrm{PSD}_n^{(i)}(f)\right],$$

where L is the number of spectral components in the [1, 70] Hz frequency range and i is the number of resulting time intervals.

3.7.3 Median Frequency (MF)

Median frequency (MF) is defined as the frequency that comprises 50% of the power and can also be calculated from $\mathrm{PSD_n}$ within a selected frequency range [26].

$$\sum_{f=1\mathrm{Hz}}^{MF^{(i)}} \mathrm{PSD}_n^{(i)}(f) = 0.5, \quad i = 1, \ldots, 32,$$

where i is the number of resulting time intervals.

3.7.4 Relative Power (RP)

Relative power (RP) is calculated by summing the contribution of the spectral components included in the conventional EEG frequency bands:

$$RP = \sum_{f \in f_p} PSD_n^{(i)}(f), \quad f_p = \{\delta, \theta, \alpha, \beta_1, \beta_2, \gamma_1, \gamma_2\},$$

where i is the number of resulting time intervals.

3.7.5 Parameter Baseline Correction

Baseline correction is carried out using the "percent change from baseline method" [7] in order to obtain a stimulus-independent characterization. Firstly, the pre-stimulus (−300 to 0 ms interval) parameter mean is subtracted from the response value for each participant (mean of the values in the [150–450] ms interval), and then the result is divided by the pre-stimulus parameter mean [26]:

$$SE = \left\langle \frac{SE|_{i\in\text{response}} - \langle SE|_{i\in\text{baseline}}\rangle}{\langle SE|_{i\in\text{baseline}}\rangle} \right\rangle,$$

where $SE|_{i\,\in\,\text{baseline}}$ and $SE|_{i\,\in\,\text{response}}$ are the pre-stimulus and evoked potential parameter means, respectively, i is the number of resulting time intervals, and $\langle\cdot\rangle$ denotes the temporal average in those baseline and response windows [12, 26].

The proportion of change with respect to baseline SE, normalized to the unit, is then used as an estimation of SE modulation (i.e., 0.05 means a 5% decrease of SE) [12].

4 Notes

1. The EEG amplifier battery cannot be charged during recording, so this step is essential.
2. The participant should be comfortable and muscularly relaxed throughout the EEG recording to minimize artifacts. The backrest should be high enough for the participant's comfort, but the head support should not press against the occipital surface of the cap. A small neck cushion may be used to provide additional separation.
3. You must be very careful not to pull on the cables, which could compromise the quality of the recording and even damage the equipment.
4. It is recommended during the preparation of the cap to ask the participant about his or her comfort, to avoid muscle tension during a recording of several minutes.
5. The impedance level cannot be correctly displayed for all electrodes (on the program display) if the impedance of ground and reference electrodes has not been reduced first.
6. Although not fundamental, we perform the following routine that helps us to lower the impedance more efficiently: select a high impedance range (e.g., 50–100 micro-ohms) and lower the impedance to the maximum displayable (green color code in the program) at all electrodes of the cap. Then perform a second run through all electrodes in a lower sensitivity range (e.g., 25–50 micro-ohms). Repeat the process up to the range 0–10 micro-ohms.

7. It is a long paradigm, so the participant may lose attention and even fall asleep. In such a case, we will warn the participant to pay attention to the test as soon as we detect this and mark the moment in the recording program.
8. The ocular artifact is usually located in the prefrontal areas (Fp1 and Fp2), while the most common muscular artifact in our work is the lateral temporal one (electrodes T7 and T8), probably linked to mandibular tension.

Acknowledgments

We appreciate the collaboration of all the participants in our research.
This work was supported by predoctoral grants from the "Consejería de Educación, Junta de Castilla y León" (Spain) and the European Social Fund (grant IDs VA-183-18 to IFL and VA-223-19 to RMBR).

References

1. Speckman E-J, Elger CE, Gorji A (2005) Neuropshysiologic basics of EEG and DC potentials. In: Niedermeyer E, da Silva FL (eds) Electroencephalography: basic principles, clinical applications, and related fields. Lippincott Williams & Wilkins
2. Sanei S, Chambers JA (2013) Event-related potentials. EEG signal processing (131–156). Wiley
3. Picton TW, Lins OG, Scherg M (1995) The recording and analysis of event-related potentials. Handb Neuropsychol 10:3–73
4. Rampil IJ (1998) A primer for EEG signal processing in anesthesia. 89:980–1002
5. Polich J (2007) Updating P300: an integrative theory of P3a and P3b. Clin Neurophysiol 118:2128–2148
6. Iglesias-Tejedor M, Díez Á, Llorca-Bofí V, et al (2022) Relation between EEG resting-state power and modulation of P300 task-related activity in theta band in schizophrenia. Prog neuro-psychopharmacology biol psychiatry 116
7. Roach BJ, Mathalon DH (2008) Event-related EEG time-frequency analysis: an overview of measures and an analysis of early gamma band phase locking in schizophrenia. Schizophr Bull 34:907–926
8. Lisman J, Buzsáki G (2008) A neural coding scheme formed by the combined function of gamma and theta oscillations. Schizophr Bull 34:974–980
9. Gomez-Pilar J, de Luis-García R, Lubeiro A et al (2018) Deficits of entropy modulation in schizophrenia are predicted by functional connectivity strength in the theta band and structural clustering. NeuroImage Clin 18:382–389
10. Winterer G, Ziller M, Dorn H et al (2000) Schizophrenia: reduced signal-to-noise ratio and impaired phase-locking during information processing. Clin Neurophysiol 111:837–849
11. Suazo V, Lubeiro A, Jurado-Barba R et al (2016) Elevated midline-parietal gamma band noise power in schizophrenia but not in bipolar patients. Eur Arch Psychiatry Clin Neurosci 266:743–753
12. Molina V, Bachiller A, Suazo V et al (2016) Noise power associated with decreased task-induced variability of brain electrical activity in schizophrenia. Eur Arch Psychiatry Clin Neurosci 266:55–61
13. Díez Á, Suazo V, Casado P et al (2014) Frontal gamma noise power and cognitive domains in schizophrenia. Psychiatry Res - Neuroimaging 221:104–113
14. Winterer G, Coppola R, Goldberg TE et al (2004) Prefrontal broadband noise, working memory, and genetic risk for schizophrenia. Am J Psychiatry 161:490–500
15. Cea-Cañas B, Díez Á, Lubeiro A et al (2021) Altered gamma band noise power in schizophrenia and bipolar patients during a cognitive task. Eur J Psychiatry 35:129–139

16. Suazo V, Díez Á, Martín C et al (2012) Elevated noise power in gamma band related to negative symptoms and memory deficit in schizophrenia. Prog Neuro-Psychopharmacology Biol Psychiatry 38:270–275
17. Gandal MJ, Edgar JC, Klook K et al (2012) Gamma synchrony: towards a translational biomarker for the treatment-resistant symptoms of schizophrenia. Neuropharmacology 62: 1504–1518
18. Uhlhaas PJ, Singer W (2015) Oscillations and neuronal dynamics in schizophrenia: the search for basic symptoms and translational opportunities. Biol Psychiatry 77:1001–1009
19. Shannon CE (1948) A mathematical theory of communication. Bell Syst tech J 27:379–423
20. Inouye T, Shinosaki K, Sakamoto H et al (1991) Quantification of EEG irregularity by use of the entropy of the power spectrum. Electroencephalogr Clin Neurophysiol 79: 204–210
21. Scheeringa R, Fries P, Petersson KM et al (2011) Neuronal dynamics underlying high- and low-frequency EEG oscillations contribute independently to the human BOLD signal. Neuron 69:572–583
22. Fernández-Linsenbarth I, Planchuelo-Gómez Á, Díez Á et al (2021) Neurobiological underpinnings of cognitive subtypes in psychoses: a cross-diagnostic cluster analysis. Schizophr Res 229:102–111
23. Von Stein A, Chiang C, König P (2000) Top-down processing mediated by interareal synchronization. Proc Natl Acad Sci U S A 97:14748–14753
24. Womelsdorf T, Schoffelen JM, Oostenveld R, et al (2007) Modulation of neuronal interactions through neuronal synchronization. Science (80-) 316:1609–1612
25. Kopell N, Ermentrout GB, Whittington MA et al (2000) Gamma rhythms and beta rhythms have different synchronization properties. Proc Natl Acad Sci U S A 97:1867–1872
26. Bachiller A, Díez A, Suazo V et al (2014) Decreased spectral entropy modulation in patients with schizophrenia during a P300 task. Eur Arch Psychiatry Clin Neurosci 264: 533–543
27. Molina V, Bachiller A, Gomez-Pilar J et al (2018) Deficit of entropy modulation of the EEG in schizophrenia associated to cognitive performance and symptoms. A replication study. Schizophr Res 195:334–342
28. Molina V, Lubeiro A, de Luis Garcia R et al (2020) Deficits of entropy modulation of the EEG: a biomarker for altered function in schizophrenia and bipolar disorder? J Psychiatry Neurosci 45:322–333
29. Delorme A, Makeig S (2004) EEGLAB: an open source toolbox for analysis of single-trial EEG dynamics including independent component analysis. J Neurosci Methods 134:9–21
30. Oostenveld R, Fries P, Maris E, et al (2011) FieldTrip: open source software for advanced analysis of MEG, EEG, and invasive electrophysiological data. Comput Intell Neurosci 2011
31. Gomez-Pilar J, de Luis-García R, Lubeiro A et al (2018) Relations between structural and EEG-based graph metrics in healthy controls and schizophrenia patients. Hum Brain Mapp 39:3152–3165
32. Möcks J, Gasser T, Köhler W (1988) Basic statistical parameters of event-related potentials. J Psychophysiol 2:61–70
33. Cea-Cañas B, Gomez-Pilar J, Núñez P et al (2020) Connectivity strength of the EEG functional network in schizophrenia and bipolar disorder. Prog Neuro-Psychopharmacology Biol Psychiatry 98:109801
34. Díez A, Suazo V, Casado P et al (2013) Spatial distribution and cognitive correlates of gamma noise power in schizophrenia. Psychol Med 43: 1175–1185
35. Winterer G, Ziller M, Dorn H et al (1999) Cortical activation, signal-to-noise ratio and stochastic resonance during information processing in man. Clin Neurophysiol 110:1193–1203
36. Bachiller A, Lubeiro A, Díez A, et al (2015) Decreased entropy modulation of EEG response to novelty and relevance in schizophrenia during a P300 task. 525–535

Chapter 9

Quantification of Endocannabinoids in Human Plasma

Aitor Villate, Markel San Nicolas, Oier Aizpurua-Olaizola, Maitane Olivares, Aresatz Usobiaga, and Nestor Etxebarria

Abstract

The determination of the concentration of endocannabinoids and related compounds in human plasma has become a matter of interest due to their implication in physiological processes and, thus, their possible relation with physiological conditions or illnesses. The analysis of these compounds though has to be carefully designed as they are found in very low concentrations, and some of them degrade easily once blood is collected. In this chapter, a simple method based on a liquid-liquid extraction and analysis by liquid chromatography tandem mass spectrometry (LC-MS/MS) is described to determine the concentration of eight of the most relevant endocannabinoids in plasma.

Key words Endocannabinoids, N-arachidonoylethanolamine, 2-Arachidonoylglycerol, Plasma, Liquid-liquid extraction, Liquid chromatography, Tandem mass spectrometry

1 Introduction

There is a growing interest in measuring accurately the circulating levels of *endocannabinoids and related compounds* (ERCs) owing to the role of the whole endocannabinoid system in the regulation of many physiological conditions and the links with many diseases [1, 2]. In addition to this, in clinical research, the interest in understanding the occurrence and distribution of many endocannabinoids in different pathologies and in the efficiency of the therapies to treat them require the monitoring of the circulating levels of ERCs, essentially in plasma but also in other biofluids such as urine and cerebrospinal fluid (CSF) [3].

The two most relevant endocannabinoids are anandamide, also known as *N*-arachidonoylethanolamine (AEA), and 2-arachidonoylglycerol (2-AG), broadly representing *N*-acylethanolamines and monoacylglycerols as major classes of membrane lipids and playing important roles in signaling reactions of lipid signaling molecules [2]. In Tables 1 and 2, we have gathered some of these

Leyre Urigüen and Rebeca Diez-Alarcia (eds.), *Schizophrenia: Methods and Protocols*, Methods in Molecular Biology, vol. 2687, https://doi.org/10.1007/978-1-0716-3307-6_9,

Table 1
Identification information (common name, molecular formula, monoisotopic mass, PubChem-ID and LM-ID) of N-acylethanolamines endocannabinoids and related compounds often analysed in clinical research

Common name	Molecular formula	Monoisotopic mass	PubChem-ID	LM-ID
N-arachidonoyl ethanolamine (AEA)	$\mathbf{C_{22}H_{37}NO_2}$	**347.282429**	**5281969**	**LMFA08040001**
alpha-methyl anandamide	$C_{23}H_{39}NO_2$	361.298079	5283455	LMFA08040016
alpha,alpha-dimethyl anandamide	$C_{24}H_{41}NO_2$	375.313729	5283456	LMFA08040017
N-(16,16-dimethy-5Z,8Z,11Z,14Z-docosatetraenoyl)-ethanolamine	$C_{26}H_{45}NO_2$	403.345029	5283457	LMFA08040018
Eicosanoyl-ethanolamide	$C_{22}H_{45}NO_2$	355.345029	3787294	LMFA08040038
N-(2-isopropyl-5Z,8Z,11Z,14Z-eicosatetraenoyl)-ethanolamine	$C_{25}H_{43}NO_2$	389.329379	5283458	LMFA08040019
N-(5Z,8Z,11Z,14Z-heneicosatetraenoyl)-ethanolamine	$C_{23}H_{39}NO_2$	361.298079	5283459	LMFA08040020
N-(5Z,8Z,11Z,14Z-docosatetraenoyl)-ethanolamine	$C_{24}H_{41}NO_2$	375.313729	5283460	LMFA08040021
N-(5Z,8Z,11Z,14Z-tricosatetraenoyl)-ethanolamine	$C_{25}H_{43}NO_2$	389.329379	5283461	LMFA08040022
N-(5Z,8Z,11Z,14Z-tetracosatetraenoyl)-ethanolamine	$C_{26}H_{45}NO_2$	403.345029	5283462	LMFA08040023
N-(17,17-dimethyl-5Z,8Z,11Z,14Z-docosatetraenoyl)-ethanolamine	$C_{26}H_{45}NO_2$	403.345029	5283463	LMFA08040025
N-(17,17-dimethyl-5Z,8Z,11Z,14Z-heneicosatetraenoyl)-ethanolamine	$C_{25}H_{43}NO_2$	389.329379	5283464	LMFA08040026
N-(17-methyl-5Z,8Z,11Z,14Z-docosatetraenoyl)-ethanolamine	$C_{25}H_{43}NO_2$	389.329379	5283465	LMFA08040027
Anandamide (20:2, n-6)	$C_{22}H_{41}NO_2$	351.313729	5283444	LMFA08040002
Anandamide (18:3, n-6)	$C_{20}H_{35}NO_2$	321.266779	5283445	LMFA08040003
Anandamide (18:2, n-6)	$C_{20}H_{37}NO_2$	323.282429	5283446	LMFA08040004
Anandamide (20:3, n-3)	$C_{22}H_{39}NO_2$	349.298079	5283447	LMFA08040005
Anandamide (18:4, n-3)	$C_{20}H_{33}NO_2$	319.251129	5283448	LMFA08040006
Linolenoyl ethanolamide (LEA)	$\mathbf{C_{20}H_{35}NO_2}$	**321.266779**	**5283449**	**LMFA08040007**
Eicosapentaenoyl ethanolamide (EPEA)	$\mathbf{C_{22}H_{35}NO_2}$	**345.266779**	**5283450**	**LMFA08040008**
Docosahexaenoyl ethanolamide (DHEA)	$\mathbf{C_{24}H_{37}NO_2}$	**371.282429**	**5283451**	**LMFA08040009**
Anandamide (20:1, n-9)	$C_{22}H_{43}NO_2$	353.329379	5283452	LMFA08040010

(continued)

Table 1
(continued)

Common name	Molecular formula	Monoisotopic mass	PubChem-ID	LM-ID
Dihomo-gamma-linolenoyl-ethanolamide	**$C_{22}H_{39}NO_2$**	**349.298079**	**5282272**	**LMFA08040011**
Palmitoyl ethanolamide (PEA)	**$C_{18}H_{37}NO_2$**	**299.282429**	**4671**	**LMFA08040013**
N-(5,8,11,14-eicosatetraynoyl)-ethanolamine	$C_{22}H_{29}NO_2$	339.219829	5283453	LMFA08040014
N-oleoyl ethanolamine (OEA)	**$C_{20}H_{39}NO_2$**	**325.298079**	**5283454**	**LMFA08040015**
5,6-DiHETrE- ethanolamide	$C_{22}H_{39}NO_4$	381.2879087	16061178	LMFA08040029
8,9-DiHETrE- ethanolamide	$C_{22}H_{39}NO_4$	381.2879087	16061179	LMFA08040030
14,15-DiHETrE- ethanolamide	$C_{22}H_{39}NO_4$	381.2879087	16061180	LMFA08040031
5(6)-EpETrE- ethanolamide	$C_{22}H_{37}NO_3$	363.277344	16061181	LMFA08040032
8(9)-EpETrE- ethanolamide	$C_{22}H_{37}NO_3$	363.277344	16061182	LMFA08040033
11(12)-EpETrE- ethanolamide	$C_{22}H_{37}NO_3$	363.277344	16061183	LMFA08040034
14(15)-EpETrE- ethanolamide	$C_{22}H_{37}NO_3$	363.277344	24778498	LMFA08040035
11,12-DiHETrE- ethanolamide	$C_{22}H_{39}NO_4$	381.2879087	16061184	LMFA08040036
Mead Acid- ethanolamide	$C_{22}H_{39}NO_2$	349.298079	16061185	LMFA08040037
Eicosatetraenoic acid ethanolamide (ETEA)	**$C_{22}H_{37}NO_3$**	**363.277344**	**35027640**	**LMFA08040039**
15-HETE- ethanolamide	$C_{22}H_{37}NO_3$	363.277344	44256595	LMFA08040040
Lauroyl ethanolamide	$C_{14}H_{29}NO_2$	243.219829	8899	LMFA08040041
Myristoyl ethanolamide	$C_{16}H_{33}NO_2$	271.251129	8890	LMFA08040042
Palmitoleoyl ethanolamide (POEA)	**$C_{18}H_{35}NO_2$**	**297.266779**	**9835868**	**LMFA08040043**
Pentadecanoyl ethanolamide	$C_{17}H_{35}NO_2$	285.266779	14455157	LMFA08040045
Docosatetraenoyl ethanolamide (DEA)	**$C_{24}H_{41}NO_2$**	**375.313729**	**5282273**	**LMFA08040047**
Margaroyl ethanolamide	$C_{19}H_{39}NO_2$	313.298079	14455158	LMFA08040049
Erucicoyl ethanolamide	$C_{24}H_{47}NO_2$	381.360679	6436482	LMFA08040050
N-Arachidonoyl dopamine (NADA)	**$C_{28}H_{41}NO_3$**	**439.308644**	**5282105**	**LMFA08020084**
Stearoyl ethanolamide (SEA)	**$C_{20}H_{41}NO_2$**	**327.313729**	**27902**	**LMFA08040051**
Behenoyl ethanolamide	$C_{24}H_{49}NO_2$	383.376329	3023585	LMFA08040052
Tricosanoyl ethanolamide	$C_{25}H_{51}NO_2$	397.391979	14455160	LMFA08040053
N-oleoyl dopamine (ODA)	**$C_{26}H_{43}NO_3$**	**417.324294**	**5282106**	**LMFA08020140**
O-arachidonoyl ethanolamine (O-AEA)	**$C_{22}H_{37}NO_2$**	**347.282429**	**5712057**	**LMFA00000014**

(continued)

Table 1
(continued)

Common name	Molecular formula	Monoisotopic mass	PubChem-ID	LM-ID
Oleamide (OLA)	$C_{18}H_{35}NO$	281.271865	5283387	LMFA08010004
N-arachidonylglycine (NAGly)	$C_{22}H_{35}NO_3$	361.261694	5283389	LMFA08020003
Vaccenic acid ethanolamide (VEA)	$C_{20}H_{39}NO_2$	325.298079	9883792	LMFA08040059

The most analyzed compounds are marked in bold

N-acylethanolamine and monoacylglycerol compounds, respectively, including key analytical information such as their molecular formula and monoisotopic mass, as well as their PubChem and LipidMaps (LM) identification codes (IDs). Most of the analytical methods and protocols are limited to some of the ERCs included in Tables 1 and 2, and the quantification of the circulating levels of such compounds allows understanding the distribution and metabolic pathways involved [4, 5]. This fact can be particularly of interest if a non-targeted approach is considered to tackle, for instance, the quest for potential biomarkers [6, 7]. In most of the cases, however, a targeted approach is an effective way to afford many clinical issues [8].

The analysis of targeted ERCs is featured by two main limitations. On the one hand, most compounds of interest are at very low concentrations, typically in low ng/mL range and many below 10 ng/mL, which requires a very sensitive instrumental approach and a high overall recovery in order to avoid false negatives [9, 10]. On the other hand, the features of the sample itself are key to ensure reliable concentrations. For instance, the way plasma and serum samples are obtained is key to assure reliable results as some biochemical transformations can happen (e.g., 2-AG → 1-AG isomerization) in both types of samples, and some ERCs are released when whole blood is kept at room temperature before being centrifuged. Therefore, critical pre-analytical parameters should be thoroughly considered, and best practice suggests the immediate processing of blood at low temperature and the storage at −80 °C until analysis. In addition to this, it is advisable to aliquot the original sample in several subsamples to avoid many freeze-thaw cycles that could end up in the transformation and/or degradation of targets. Finally, it is highly desirable to standardize all analytical steps to minimize the variability of the whole process; otherwise the reliability of the results may be questioned [11].

Most of the reported analytical procedures start with addition of cold acetonitrile and the subsequent centrifugation to precipitate the proteins in plasma [12]. The use of aprotic solvents (i.e., acetonitrile instead of methanol) has become a must in the analysis

Table 2
Identification information (common name, molecular formula, monoisotopic mass, PubChem-ID and LM-ID) of monoacylglycerols endocannabinoids and related compounds often analysed in clinical research

Common name	Molecular formula	Monoisotopic mass	PubChem-ID	LM_ID
MG(12:0/0:0/0:0)	$C_{15}H_{30}O_4$	274.21441	656773	LMGL01010008
MG(16:0/0:0/0:0)[rac]	$C_{19}H_{38}O_4$	330.27701	14900	LMGL01010001
MG(17:0/0:0/0:0)[rac]	$C_{20}H_{40}O_4$	344.29266	107036	LMGL01010002
MG(18:0/0:0/0:0)[rac]	$C_{21}H_{42}O_4$	358.30831	24699	LMGL01010003
MG(18:1(11E)/0:0/0:0)[rac]	$C_{21}H_{40}O_4$	356.29266	5283467	LMGL01010004
MG(18:1(9Z)/0:0/0:0)[rac]	$C_{21}H_{40}O_4$	356.29266	5283468	LMGL01010005
MG(18:2(9Z,12Z)/0:0/0:0)[rac]	$C_{21}H_{38}O_4$	354.27701	5283469	LMGL01010006
MG(20:0/0:0/0:0)[rac]	$C_{23}H_{46}O_4$	386.33961	3246944	LMGL01010007
MG(16:0/0:0/0:0)	$C_{19}H_{38}O_4$	330.27701	3084463	LMGL01010009
2-arachidonoyl glycerol (2-AG)	**$C_{23}H_{38}O_4$**	**378.27701**	**5282280**	**LMGL01010023**
1-Octylglycerol	$C_{11}H_{24}O_3$	204.172545	15956913	LMGL01020027
1-Hexylglycerol	$C_9H_{20}O_3$	176.141245	10313344	LMGL01020028
2-Hexylglycerol	$C_9H_{20}O_3$	176.141245	82535	LMGL01020029
2-Octylglycerol	$C_{11}H_{24}O_3$	204.172545	372441	LMGL01020030
2-oleoylglycerol (2-OG)	**$C_{21}H_{40}O_4$**	**356.29266**	**5319879**	**LMGL01010024**
2-palmitoylglycerol (2-PG)	$C_{19}H_{38}O_4$	330.27701	123409	LMGL01010025
MG(0:0/20:5 (5Z,8Z,11Z,14Z,17Z)/0:0)	$C_{23}H_{36}O_4$	376.26136	11660820	LMGL01010026
MG(0:0/22:6 (4Z,7Z,10Z,13Z,16Z,19Z)/0:0)	$C_{25}H_{38}O_4$	402.27701	46896303	LMGL01010027
Noladin Ether (NE)	**$C_{23}H_{40}O_3$**	**364.297745**	**6483057**	**LMGL01020034**
15-HETE-G	$C_{23}H_{38}O_5$	394.271924	10200684	LMGL01010030
2-arachidonoylglycerol ether (2-AGE)	**$C_{23}H_{36}O_5$**	392.256274	129715582	
1-arachidonoyl glycerol (1-AG)	**$C_{23}H_{38}O_4$**	**378.277010**	**114672**	**LMGL01010032**
2-linoleoylglycerol (2-LG)	**$C_{21}H_{38}O_4$**	**354.277010**	**5365676**	**LMGL01010033**

The most analyzed compounds are marked in bold

of ERCs to minimize the abovementioned isomerization step (2-AG → 1-AG) [13]. The supernatant liquid is transferred to another vial, and then the extract can be evaporated to increase the concentration before injection in the chromatographic setup. Although it is not always necessary, in order to eliminate potential interfering compounds, an additional cleanup step, primarily using solid phase extraction, could be performed before the pre-concentration step [12]. The analysis of targets compiled in Tables 1 and 2 is typically performed in a high-performance liquid chromatography–tandem mass spectrometry (HPLC-MS/MS) chromatographic setup using reverse phase conditions, whereas the use of high-resolution mass spectrometry (HRMS) is required for the identification of unknown biomarkers.

The analytical protocol scheme shown in this work has been validated to analyze eight endogenous compounds in plasma including the mentioned AEA, 2-AG, and 1-AG, and five of the most relevant *N*-acylethanolamines: palmitoyl ethanolamide (PEA), linoleoyl ethanolamide (LEA), oleoyl ethanolamide (OEA), docosatetraenoyl ethanolamide (DEA), and *N*-arachidonoyl dopamine (NADA). However, as explained hereinafter, this protocol can be extended to more compounds of the same family if they are considered as potential target compounds. Briefly, the described procedure consists on a liquid-liquid extraction supported by a bead mill homogenizer and a subsequent pre-concentration of the organic phase prior to analyze the targets by LC-MS/MS. The high extraction efficiency and minimal matrix effect shown by this protocol render accurate results in the quantification of the mentioned endogenous compounds. Moreover, the good method throughput time eases to run up to 70 samples per day, which allows the analysis of a huge number of samples typical in clinical assays.

2 Materials

2.1 Supplies

1. Adjustable micropipettes: 2–20 μL, 20–200 μL, 100–1,000 μL, and 500–5,000 μL.
2. 2 mL screw top amber borosilicate glass vials, with corresponding screw caps and silicone septa.
3. 4 mL screw top amber borosilicate glass vials, with corresponding PTFE lined storage screw caps.
4. 1,000 mL volumetric flasks.
5. 1,000 mL glass reagent bottles for HPLC mobile phases.
6. BD Vacutainer® Push Bottom Blood Collection Set with 21-gauge butterfly needle and 12″ tubing.
7. 6 mL BD Vacutainer® glass tubes, containing 1 mL of Acid Citrate Dextrose (ACD) solution A: trisodium citrate (22.0 g/L), citric acid (8.0 g/L), and dextrose (24.5 g/L).

8. Soda glass Pasteur pipettes.
9. 1.5 mL polypropylene centrifuge tubes.
10. 2 mL homogenization polypropylene tubes.
11. 1 mL polypropylene syringes.
12. 0.22 μm polypropylene syringe filters.
13. 250 μL conical glass vial inserts (5.6 × 31 mm).

2.2 Equipment

1. Analytical balance with a precision of ±0.1 mg.
2. Centrifuge for blood samples (Eppendorf 5810R or similar) (*see* **Note 1**).
3. Centrifuge for extractions (Allegra X-30R or similar) (*see* **Note 1**).
4. Bead mill homogenizer (Precellys® 24 coupled to Cryolys® or similar) (*see* **Note 2**).
5. Nitrogen evaporator (Turbovap or similar).
6. Vortexer.
7. LC-MS/MS, e.g., 1290 Infinity II liquid chromatograph coupled to a 6430 Triple Quad LC/MS mass spectrometer, with an Edwards 28 E2M28 vacuum pump and a RMSN-2 big universal trap for nitrogen.
8. Analytical column: Kinetex® C18 column 15 x 3 mm, 2.6 μm.
9. Analytical pre-column: SecurityGuard™ ULTRA Cartridges UHPLC C18 2 × 3 mm.
10. In-line filter: 1290 Infinity II in-line filter, 0.3 μm, 2 mm ID.

2.3 Reagents

1. HPLC grade methyl tert-butyl ether (MTBE) (*see* **Note 3**).
2. HPLC grade acetonitrile (ACN) (*see* **Note 4**).
3. HPLC grade formic acid (HCOOH).
4. HPLC grade methanol (MeOH).
5. HPLC grade water (H_2O).
6. Solid standards of the measured endocannabinoids: 1-AG, 2-AG, AEA, PEA, LEA, OEA, DEA, and NADA (*see* **Note 5**).
7. Solid standards of the following deuterated endocannabinoids: 2-Arachidonoyl glycerol-d8 (2-AG-d8), arachidonoyl ethanolamide-d8 (AEA-d8), palmitoyl ethanolamide-d4 (PEA-d4), and *N*-arachidonoyl dopamine-d8 (NADA-d8).
8. Simulated plasma fluid (SPF) (*see* **Note 6**).
9. Nitrogen gas for evaporator (recommended purity ≥99.999%).
10. Nitrogen gas for LC-MS/MS (purity ≥99.999%).
11. Liquid nitrogen.

2.4 Standard Solutions

1. Individual endocannabinoid stock solutions of 2,500 μg/mL: Weight 5 mg of each of the solid endocannabinoid standards in 4 mL amber glass vials. Add 2 mL of ACN and cap and dissolve by vortexing to obtain individual stock solutions (*see* **Note 7**).
2. Endocannabinoid mix solution of 100 μg/mL (Working Solution 1, WS1): Add 60 μL of each of the individual endocannabinoids 2,500 μg/mL stock solution in a 2 mL amber glass vial, and dilute the mix by adding 1,000 μL of ACN. Cap and vortex vigorously (*see* **Note 8**).
3. Endocannabinoid mix solution of 2.5 μg/mL (WS2): Add 25 μL from WS1 and 975 μL of ACN in a 2 mL amber glass vial. Cap and vortex vigorously.
4. Individual deuterated endocannabinoid stock solutions of 50 μg/mL: Weight 100 μg of each of the solid endocannabinoid standards in 4 mL amber glass vials. Add 2 mL of ACN, and cap and dissolve by vortexing to obtain individual stock solutions.
5. Internal standard solution: Prepare a stock solution with the deuterated endocannabinoids in a concentration of 1.25 μg/mL in a 4 mL amber glass vial. Add 100 μL from each of the individual deuterated endocannabinoid stock solutions of 50 μg/mL, and then add 3,600 μL of ACN. Cap and vortex vigorously.
6. Mix solution of 1,000 ng/mL (WS3): Add 40 μL from the WS1 and 80 μL from each of the individual deuterated endocannabinoid stock solutions of 50 μg/mL and 3,640 μL of ACN in a 4 mL amber glass vial to obtain a mix solution with all the measured endocannabinoids and deuterated endocannabinoids. Cap and vortex vigorously.
7. Mix solution of 100 ng/mL (WS4): Add 400 μL from the WS3 and 3,600 μL of ACN in a 4 mL amber glass vial. Cap and vortex vigorously.
8. Mix solution of 1 ng/mL (WS5): Add 40 μL from the WS4 and 3,960 μL in a 4 mL amber glass vial. Cap and vortex vigorously.

2.5 Calibration Solutions (See Note 9)

1. Blank. Transfer 1,000 μL of ACN in a 1.5 mL amber glass vial.
2. Calibration solution 1 (0.5 ng/mL): Add 500 μL from WS5 and 500 μL of ACN in a 1.5 mL amber glass vial. Cap and vortex vigorously.
3. Calibration solution 2 (1 ng/mL): Transfer 1,000 μL from WS5 to a 1.5 mL amber glass vial. Cap and vortex vigorously.
4. Calibration solution 3 (5 ng/mL): Add 50 μL from the WS4 and 950 μL of ACN in a 1.5 mL amber glass vial. Cap and vortex vigorously.

5. Calibration solution 4 (10 ng/mL): Add 100 μL from the WS4 and 900 μL of ACN in a 1.5 mL amber glass vial. Cap and vortex vigorously.
6. Calibration solution 5 (25 ng/mL): Add 250 μL from the WS4 and 750 μL of ACN in a 1.5 mL amber glass vial. Cap and vortex vigorously.
7. Calibration solution 6 (50 ng/mL): Add 10 μL from the WS4 and 990 μL of ACN in a 1.5 mL amber glass vial. Cap and vortex vigorously.
8. Calibration solution 7 (100 ng/mL): Transfer 1,000 μL of the WS4 to a 1.5 mL amber glass vial. Cap and vortex vigorously.
9. Calibration solution 8 (200 ng/mL): Add 200 μL from the WS3 and 800 μL of ACN in a 1.5 mL amber glass vial. Cap and vortex vigorously.
10. Calibration solution 9 (300 ng/mL): Add 300 μL from the WS3 and 700 μL of ACN in a 1.5 mL amber glass vial. Cap and vortex vigorously.

2.6 Quality Control Solutions

1. Low concentration quality control solution (LQC): Add 100 μL from each of the individual deuterated endocannabinoid stock solutions of 50 μg/mL, 20 μL from WS2, and 3,580 μL of ACN. Cap and vortex vigorously.
2. Medium concentration quality control solution (MQC): Add 100 μL from each of the individual deuterated endocannabinoid stock solutions of 50 μg/mL, 200 μL from WS2, and 3,400 μL of ACN. Cap and vortex vigorously.
3. High concentration quality control solution (HQC): Add 100 μL from each of the individual deuterated endocannabinoid stock solutions of 50 μg/mL, 50 μL from the WS1, and 3,550 μL of ACN. Cap and vortex vigorously.

2.7 Mobile Phases

1. Mobile phase A: 0.1% HCOOH in H_2O. Fill approximately the half of a 1,000 mL volumetric flask with LC-MS quality H_2O and add to it 1,000 μL of HCOOH. Fill to 1,000 mL with H_2O and cap and mix by inversion. Transfer the content to a 1,000 mL glass reagent bottle for HPLC mobile phases and place it into the HPLC.
2. Mobile phase B: 0.1% HCOOH in MeOH. Fill approximately the half of a 1,000 mL volumetric flask with MeOH and add to it 1,000 μL of HCOOH. Fill to 1,000 mL with MeOH and cap and mix by inversion. Transfer the content to a 1,000 mL glass reagent bottle for HPLC mobile phases and place it into the HPLC.

3 Methods

3.1 Blood Collection and Plasma Extraction

1. Collect blood by venepuncture using the BD Vacutainer® push bottom blood collection set with 21-gauge butterfly needle and 12″ tubing connected to the 6 mL BD Vacutainer® collection tubes containing ACD solution A (*see* **Note 10**).
2. Once approximately 5 mL of blood are collected, mix it with the ACD solution by inverting the collection tubes 3–4 times (*see* **Note 11**).
3. Immediately (as soon as possible), centrifuge the blood samples at 2,000 × *g* at 4 °C.
4. Immediately after centrifugation, remove the supernatant with a Pasteur pipette and place it into 1.5 mL polypropylene tubes (*see* **Note 12**).
5. Store the plasma samples at −80 °C immediately (as soon as possible) after collection (*see* **Note 13**).

3.2 Extraction of Endocannabinoids from Plasma

1. Thaw slowly the plasma samples by storing them at 4 °C (*see* **Note 14**).
2. When the sample is completely thawed, vortex vigorously to homogenize its content.
3. For each procedural replicate, take 500 μL of the homogenized sample and place them into 2 mL homogenization tubes. At least three procedural replicates are recommended for each individual (*see* **Note 15**).
4. In order to correct potential losses that can happen during the sample preparation process, add to each sample 20 μL of the internal standard solution and cap and vortex vigorously to mix it with the plasma (*see* **Note 16**).
5. Add 1,000 μL of precooled MTBE (4 °C) to the homogenization tubes and cap and place them into the bead mill homogenizer, once the homogenizer has reached 4 °C.
6. Homogenize the samples using three cycles of 20 s at 6,500 rpm and 20 s of rest between cycles. Maintain the temperature at 4 °C in the whole homogenization process.
7. Place the homogenization tubes in the centrifuge and centrifuge the samples at 13,000 × *g* for 5 min at 4 °C.
8. After centrifugation, the organic and the aqueous phase will be clearly distinguishable. Collect the upper phase (organic phase) with a Pasteur pipette and place the collected phase in a 2 mL amber glass vial (*see* **Note 17**).
9. Place the amber glass vials with the organic phases in the nitrogen evaporator (preheated at 35 °C) and evaporate to dryness with nitrogenous gas with a pressure of 5 psi (*see* **Note 18**).

10. Add 200 µL of ACN to the dried samples and cap and vortex vigorously to assure that the analytes are dissolved in the new solvent (*see* **Note 19**).
11. Filter the final solutions with syringe and disk filter and place them into 250 µL conical inserts. To do this, take the content from the amber glass vials with a Pasteur pipette and place it into syringes coupled to a 0.22 µm disk filters. Apply pressure to the syringes and collect the filtered samples in 250 µL inserts directly. Put the inserts in 2 mL amber glass vials and cap them.
12. If immediate analysis is not possible, store the sample extracts at −80 °C until analysis (*see* **Note 20**).

3.3 Analysis with LC-MS/MS

1. Analyze the filtered samples in the 1290 Infinity II liquid chromatograph coupled to a 6430 Triple Quad LC/MS, using a Kinetex® C18 column 15 x 3 mm, 2.6 µm column to separate the analytes, a SecurityGuard™ ULTRA Cartridges UHPLC C18 2 x 3 mm before the column to protect the column, and a 1290 Infinity II in-line filter to protect the chromatograph.
2. Injection volume: 10 µL (*see* **Note 21**).
3. Column temperature: 35 °C.
4. Flow rate: 0.3 mL/min.
5. The chromatographic method starts with a 70% of mobile phase B. It is raised to 90% of mobile phase B in the first 2 min, and this composition is maintained for 5 min. Then, mobile phase B is raised to 100% in 2 min, and it is maintained there for 4 min. Finally, mobile phase B is lowered to 70% in 2 min and it is maintained there for other 2 min to reach initial conditions (*see* **Note 22**).
6. The conditions used in the electrospray ionization source of the mass spectrometer are detailed in Table 3 (*see* **Note 23**).
7. Multiple reaction monitoring (MRM) conditions, including precursor ion, product ions, collision energies, and fragmentor voltages are detailed in Table 4. The retention times of the compounds separated at described conditions are also defined in Table 4. The transition to the first product ion is the one used for quantification and the second one for confirmation purposes except PEA and PEA-d4, where the second transition is used (*see* **Note 24**).
8. Build an external calibration using the values obtained from the integration of the chromatographic signals of the calibration solutions for each of the measured compounds (*see* **Note 25**).
9. Identify the targeted compounds in the chromatograms of the final extracts taking into account the retention times and the two MRM transitions. Once identified, integrate the corresponding chromatographic peak (*see* **Note 26**).

Table 3
Mass spectrometer parameters

Parameter	Setting
Ion source	Electrospray ionization (ESI)
Polarity	Positive
Time filtering (Peak Width)	0.07 min
Scan Type	Dynamic MRM
Delta EMV (+)	200
Cycle Time	500 ms
Drying, nebulisation and collision gas	Nitrogen
Gas Temperature	350 °~C
Gas Flow	11 L/min
Nebuliser	30 psi
Capillary Voltage	5,500 V
Cell Accelerator Voltage	4 V

10. Obtain the concentration of the targets in the final extracts by interpolation using the obtained integration values for each compound and its corresponding calibration curve equation.
11. Express the results giving the concentration of each of the analytes in ng/mL in plasma (*see* **Note 27**).

3.4 Procedural Blanks and Quality Controls

1. Procedural blanks: Take 500 μL of SPF for each procedural replicate and place them into 2 mL homogenization tube. Process the sample following the procedure of 3.2 (after the third step) and 3.3 subheadings. Use the results to check that there is no source of contamination of the analytes in the analytical procedure. At least three procedural replicates are recommended.
2. Quality controls: Take 500 μL of SPF for each procedural replicate and place them into 2 mL homogenization tubes. Spike them in low, medium, and high analyte levels by adding 20 μL from the LQC, MQC, and HQC solutions, respectively. Process the samples following the procedure of 3.2 (after the fifth step) and 3.3 subheadings. Use the results to check that the trueness and the precision of the method meet the required acceptance criteria for all the analytes in the three levels. At least three procedural replicates are recommended for each level (*see* **Note 28**).

Table 4
MRM conditions and retention times for the measured compounds

Compound	Precursor ion (*m/z*)	Product ion (*m/z*)	Collision energy (eV)	Fragmentor voltage (V)	Retention Time (min)
1-AG	379.3	287.2 269.2	9 13	134	7.5
2-AG	379.3	287.2 269.2	9 9	118	7.2
AEA	348.3	91.0 81.1	45 29	118	7.0
DEA	376.3	79.0 91.0	45 45	134	8.7
LEA	324.3	81.1 95.1	37 29	118	7.0
NADA	440.3	137.1 154.1	25 9	150	6.9
OEA	326.3	309.3 83.1	9 25	134	8.4
PEA	300.3	283.3 71.1	9 21	134	7.8
2-AG-d8	387.3	294.3 293.3	9 9	118	7.2
AEA-d8	356.3	70.1 82.1	45 45	118	7.0
NADA-d8	448.4	137.1 138.1	29 29	150	6.9
PEA-d4	304.3	287.3 72.2	9 21	118	7.8

4 Notes

1. Centrifuge for blood samples should be able to centrifuge at 2,000 × *g* and to work at low temperatures (at least 4 °C) to avoid changes in analyte composition. Centrifuge for extracted samples should be able to centrifuge at 13,000 × *g* and to work at low temperatures (at least 4 °C) too. It is highly recommended to precool the centrifuges at the desired temperature before centrifugation to ensure the proper temperature is reached.
2. Bead mill homogenizer should be able to work at low temperatures (at least 4 °C). Other homogenizers can be used, as long as the chosen homogenization technique does not increase the temperature of the samples and does not take too much time. A

good alternative is simply to vortex the samples for 1–2 min; it is short process and it is not very energetic so it will not increase the temperature of the samples considerably. If other homogenization techniques are used, skip steps 5 and 6 from the 3.2 subheading and adapt the protocol to the chosen technique.

3. MTBE is the chosen extraction solvent, as it showed the best recoveries and repeatability for the purposed extraction procedure compared to other ones. Other common solvents used for endocannabinoid extraction include toluene and ethyl acetate, which can be good alternatives in case of not having MTBE. In any case, the chosen solvent should be an aprotic solvent of low polarity, considering that endocannabinoids degrade easily in protic solvents (such as alcohols or water), and therefore, their use is not recommended. It is also suggested to use solvents with low boiling temperatures to ease the evaporation step (*see* **Note 18**).
4. ACN is the chosen solvent for standard preparation and for the reconstitution of the final extract. Other aprotic solvents could be used for this purpose, but when it comes to the analysis of endocannabinoids, ACN is the preferred choice.
5. As mentioned in the introduction, these are the proposed eight targeted compounds, which include the most relevant and usual endocannabinoids. However, the procedure can be extended to more targeted compounds. Most of the standards of these compounds are commercially available as solids.
6. Analyte-free plasma sample can be used as SPF and it is used for the procedural blanks and quality control experiments. If standard SPF is not available, it can be prepared by diluting the right concentrations of electrolytes in water [14].
7. Most of the solid standards will dissolve in ACN by just shaking and vortexing the vials. If a heterogeneous solution is not reached after this, ultrasonication may help. Do not ever use heat to help solubilization as standards could be damaged. Among the present standards, PEA has the poorest solubility in ACN and may not be completely dissolved. To help solubilization, 100 μL of chloroform can be added. If chloroform is added, take it into account to recalculate its real concentration. It is highly recommended to weight all the prepared solutions to determine the nominal concentration of the endocannabinoids, as volumes should not be considered addable for the calculation of concentrations.
8. If further standards are included in the analysis, prepare 2,500 μg/mL stock solutions from the solid standards and include them in the last mix solution.
9. Calibration range may vary depending on the expected endocannabinoid levels and the detection limits of the used LC-MS/MS. The proposed calibration rank tries to cover the

Table 5
Reported endocannabinoid concentration ranges in plasma

Analyte	Concentration in plasma (ng/mL)	References
AEA	0.2–5	[13]
1-AG	–	[13]
2-AG	0.6–15	[13]
OEA	1.2–53	[13]
PEA	1.4–125	[13]
LEA	1.2–18.0	[10]
DEA	<1	[9, 15]
NADA	<1	[9, 15]

full possible ranges of endocannabinoid levels in plasma, taking into account what has been reported in the bibliography (Table 5). Additionally, the lower level of the shown calibration range is limited by the detection limits of the used LC-MS/MS. Before establishing the lower level of the calibration range, it is recommended to check the detection limit of the instrument and fix it. This can be crucial, as some of the analytes could be around detection limit levels in the final extract. In order to determine the instrumental quantification limit, a good option is to prepare lowering concentration solutions of the target compounds and check which is the lowest measurable concentration for each of the analytes. Briefly, prepare an external calibration of the target compounds in a very low concentration range (i.e., 0.01–10 ng/mL range in ACN solvent). Inject those low calibration points in triplicate to calculate instrumental limit of quantifications, which are defined for each analyte as the lowest concentration level that fitted the calibration curve with a relative standard deviation (RSD %) and systematic error lower than 30%. See **Notes 19** and **21** to see other topics related to detection limits.

10. Take special care in safety elements as blood, biological specimens, and materials that are in contact with them are considered biohazards. Use personal protective equipment (i.e., coat, gloves, and safety glasses) and after finishing the analysis discard all blood collection materials in biohazard containers approved for this purpose. Simultaneously, use aseptic techniques and sterile disposables to prevent contamination of blood. Use individual needles, blood collection sets, and tubes for each of the individuals to avoid transmission of possible infections between individuals and avoid sample cross-

contamination. Nevertheless, only personnel trained in venepuncture should perform this step.

11. Collect the specified blood volume to ensure that the blood/anticoagulant ratio is appropriate. The BD Vacutainer® push bottom blood collection set is prepared to stop drawing blood once the tube is completely filled. If a tube is partially filled, consider avoiding this sample as blood/anticoagulant ratio may not be suitable.
12. Pay attention to the color of the plasma samples, as its color can be an indicator of its suitability. Yellowish plasma indicates that the sample is suitable for analysis, whereas orange-reddish color plasma indicates that surely hemolysis has happened during the blood collection step and it should be discarded for analysis. 21-gauge needles are recommended to minimize hemolysis.
13. It is crucial to maintain the cold chain of the fresh plasma samples and to store them as soon as possible at the indicated temperature to avoid changes in the endocannabinoid levels. Endocannabinoids are especially unstable when they are in contact with protic solvents, such as water present in plasma. In consequence, their levels can change drastically in a few hours if they are stored at room temperature and even if they are stored in the fridge (4 °C), especially in the case of 1-AG and 2-AG, due to 2-AG to 1-AG isomerization. Moreover, freeze-thaw cycles can contribute to changes in analyte composition, so avoid doing it more than once. It is not recommended to store the plasma samples more than 3 months [9, 13].
14. They can also be thawed at room temperature, but in order to guarantee the preservation of endocannabinoids, slow thawing at 4 °C is recommended. This slow thawing process can last 3–4 h.
15. 500 μL is the suggested volume for this procedure, but it could vary depending on the needs and limitations of each case. The detailed procedure can be used for smaller sample volumes, but all sample/solvent ratios should be maintained or adjusted to the used sample volume.
16. Isotopically marked compounds (deuterated analogues in this case) have the same extraction recovery and spectrometric behavior as their counterparts so they are ideal compounds to be used as surrogate compounds or internal standards when MS is used as detection analytical technique. The comparison of the nominal concentration and the experimental concentration of the deuterated analogues provides the information about the recovery, which is used to correct the concentration of the target compounds and provide accurate results in terms

Table 6
Isotopically marked compounds used to correct the concentration of targets

Deuterated compound	Corrected analyte
2-AG-d8	2-AG
AEA-d8	AEA, LEA
PEA-d4	1-AG, DEA, OEA, PEA
NADA-d8	NADA

of trueness and precision. Table 6 resumes which deuterated analogue should be used to correct the recovery of targets.

If more than the proposed targeted compounds are included in the analysis, their respective deuterated compounds should be included too. In some cases, there is no need to include a deuterated counterpart for each analyte (as shown above), but this kind of compatibility should be checked either in literature or with spiked blank samples/quality control samples.

17. Try to recover the whole phase quantitatively and avoid taking anything from the aqueous phase, as it may compromise the drying step. Internal standard solution will correct the losses if the whole phase is not recovered. However, if there is no chance to add the deuterated compound solution, in order to get adequate precision, it is recommended to take a certain volume that can be easily pipetted with an adjustable micropipette, such as 700 μL. If a solvent with higher density than water is used for liquid-liquid extraction purposes, the collection of that phase could be hindered.
18. Keep an eye on the amber glasses when turning on the nitrogen flux to ensure that no losses occur due to splashing. This effect can be avoided if the nitrogen pressure is arisen slowly. The average evaporation time in the defined conditions is around 5 min. In order to prevent analyte loss during the evaporation step, large evaporation times should be avoided. Boiling temperature of the organic solvent should be also considered if a solvent different to MTBE is used. For this reason, ethyl acetate may be a better option compared to toluene, whose boiling temperature is almost the double of MTBE's.
19. In the case of having issues related to detection limits, reconstitution volumes can be lowered to 100 μL or even 50 μL to increase analytes' concentration in the final extract. However, this could compromise and/or hinder the filtration step. If lower reconstitution volumes are used, small volume inserts should be used in order to enable the analysis.

20. Once analytes are dissolved in ACN, samples can be stored at − 20 °C until analysis for 2 weeks.
21. Injection volume may vary depending on the used chromatographic setup, column, and detection system, which all affect the instrumental limits of detection. Follow always the recommendations of the used column's manufacturer to never exceed the column's maximum injection volume. Higher injection volumes can help in the improvement of the detection limits, but too high injection volumes can also lead to the widening of chromatographic peaks. To find an appropriate injection volume, it is recommended to check a few injection volumes (e.g., 2 μL, 3 μL, 5 μL, 10 μL) with one of the standard solutions and evaluate each volumes' suitability based on peak shape and width of the chromatographic peaks (i.e., narrow Gaussian peak).
22. Gradient conditions should be optimized if a different column, mobile phase, and flow are used or if more analytes are added to the analysis.
23. The measurement conditions for the ionization source of the mass spectrometer should be optimized if a different device is used.
24. If a different device is used, the collision energies and fragmentor voltages of all transitions should be optimized. If more analytes are analyzed, precursor and product ions have to be defined, and their corresponding collision energies and fragmentor voltages optimized. In different analysis batch, check and fix the retention times of the targets using standard solutions since the retention times can slightly be shifted even if the same LC and column are used.
25. A simple lineal regression between each compound's concentration at different calibration points and their corresponding peak integration values will work for the calibration building.
26. For the correct identification and integration of chromatographic peaks of each analyte, discard peaks whose retention time is shifted more than 0.2 min from its corresponding retention time and avoid integrating chromatographic peaks defined with less than five points.
27. Calculate the concentration in plasma dividing the concentration of the final extract by 2.5 (this calculation will only be applicable if the exact same procedure has been followed). Finally, correct the concentrations taking into account the results of the deuterated endocannabinoids (*see* **Note 16**).
28. Quality control samples should cover different concentration ranges to check that the analytical method works in all of them. Typically, three levels are checked; in this case, quality control

samples are spiked to result in concentrations of 0.5 ng/mL (lower level, near detection limit), 5 ng/mL (medium level), and 50 ng/mL (higher level) in SPF.

Trueness of the analysis is calculated by dividing the obtained analyte concentration values by the expected concentration in each of the tested levels, and they are given in percentages. Trueness values between 80% and 120% are acceptable.

Precision of the analysis is evaluated taking into account the obtained coefficient variation % (CV %) in each level. This is calculated by dividing the obtained variance (from the analyte concentration values of the replicates) by the obtained analyte concentration, and they are given in percentages. CV % values below 20% are acceptable. In the case of routine analysis, inter-day precision can be checked as well, by comparing the results of samples analyzed in different days [16].

Quality control samples can be used as well when large amounts of samples are analyzed to check that analyte composition does not change along the sequence. This can be done by analyzing one of the quality control samples several times along the sequence and by comparing the chromatographic peak areas.

References

1. Watson JE, Kim JS, Das A (2019) Emerging class of omega-3 fatty acid endocannabinoids & their derivatives. Prostaglandins Other Lipid Mediat 143:106337. https://doi.org/10.1016/j.prostaglandins.2019.106337
2. Tsuboi K, Uyama T, Okamoto Y et al (2018) Endocannabinoids and related N-acylethanolamines: biological activities and metabolism. Inflammation and Regeneration 38:28. https://doi.org/10.1186/s41232-018-0086-5
3. Aizpurua-Olaizola O, Elezgarai I, Rico-Barrio I et al (2017) Targeting the endocannabinoid system: future therapeutic strategies. Drug Discov Today 22(1):105–110. https://doi.org/10.1016/j.drudis.2016.08.005
4. Joaquim HPG, Costa AC, Pereira CAC et al (2022) Plasmatic endocannabinoids are decreased in subjects with ultra-high risk of psychosis. Eur J Neurosci 55:1079–1087. https://doi.org/10.1111/ejn.15509
5. Potvin S, Mahrouche L, Assaf R et al (2020) Peripheral endogenous cannabinoid levels are increased in schizophrenia patients evaluated in a psychiatric emergency setting. Front Psych 11:628. https://doi.org/10.3389/fpsyt.2020.00628
6. Navarrete F, García-Gutiérrez MS, Jurado-Barba R et al (2020) Endocannabinoid system components as potential biomarkers in psychiatry. Front Psych 11:315. https://doi.org/10.3389/fpsyt.2020.00315
7. Berman P, Sulimani L, Gelfand A et al (2020) Cannabinoidomics – an analytical approach to understand the effect of medical Cannabis treatment on the endocannabinoid metabolome. Talanta 219:121336. https://doi.org/10.1016/j.talanta.2020.121336
8. Seabra G, Falvella ACB, Guest PC et al (2018) Proteomics and lipidomics in the elucidation of endocannabinoid signaling in healthy and schizophrenia brains. Proteomics 18:1700270. https://doi.org/10.1002/pmic.201700270
9. Sempio C, Klawitter J, Jackson M et al (2021) Analysis of 14 endocannabinoids and endocannabinoid congeners in human plasma using column switching high-performance atmospheric pressure chemical ionization liquid chromatography–mass spectrometry. Anal Bioanal Chem 413(12):3381–3392. https://doi.org/10.1007/s00216-021-03280-0
10. Röhrig W, Achenbach S, Deutsch B et al (2019) Quantification of 24 circulating endocannabinoids, endocannabinoid-related

compounds, and their phospholipid precursors in human plasma by UHPLC-MS/MS. J Lipid Res 60(8):1475–1488. https://doi.org/10.1194/jlr.D094680

11. Kratz D, Sens A, Schäfer SMG et al (2022) Pre-analytical challenges for the quantification of endocannabinoids in human serum. J Chromatogr B Anal Technol Biomed Life Sci 1190: 123102. https://doi.org/10.1016/j.jchromb.2022.123102
12. Shafeeque A, Kalra H, Gupta A et al (2012) HybridSPE: a novel technique to reduce phospholipid-based matrix effect in LC-ESI-MS Bioanalysis. J Pharm BioAllied Sci 4(4): 267–275. https://doi.org/10.4103/0975-7406.103234
13. Gurke R, Thomas D, Schreiber Y et al (2019) Determination of endocannabinoids and endocannabinoid-like substances in human K3EDTA plasma – LC-MS/MS method validation and pre-analytical characteristics. Talanta 204:386–394. https://doi.org/10.1016/j.talanta.2019.06.004
14. Oyane A, Kim H-M, Furuya T et al (2003) Preparation and assessment of revised simulated body fluids. J Biomed Mater Res 65A:188–195. https://doi.org/10.1002/jbm.a.10482
15. Hauer D, Schelling G, Gola H et al (2013) Plasma concentrations of endocannabinoids and related primary fatty acid amides in patients with post-traumatic stress disorder. PLoS One 8(5):e62741. https://doi.org/10.1371/journal.pone.0062741
16. González O, Blanco ME, Iriarte G et al (2014) Bioanalytical chromatographic method validation according to current regulations, with a special focus on the non-well defined parameters limit of quantification, robustness and matrix effect. J Chromatogr A 1353:10–27. https://doi.org/10.1016/j.chroma.2014.03.077

Chapter 10

Neuropsychological and Functional Assessment of Cognitive Impairments in Schizophrenia

María Recio-Barbero, Noemi Losada-León, Oihane Mentxaka, Janire Cabezas-Garduño, and Rafael Segarra

Abstract

Cognitive deficits are a central feature of psychotic disorders. Their impact and relevance for clinical prognosis and functional recovery, together with their usefulness in the development of novel therapeutic targets, have emphasized the role of cognition in the diagnosis and therapeutic management of schizophrenia. Here we describe the main aspects to consider before, during, and after the neuropsychological assessment of main cognitive domains affected in schizophrenia, from a research perspective toward clinical practice.

Key words Cognition, Cognitive dysfunction, Functionality, Assessment, First-episode psychosis, Schizophrenia

1 Introduction

Cognitive deficits are considered a core feature of psychotic disorders. Subtle cognitive changes have been identified prior to the onset of the illness [1, 2]. These changes cannot be explained by the presence of other clinical symptoms or treatment side effects and have been widely described in first-degree relatives of individuals affected by a first-episode psychosis (FEP) [3]. Furthermore, the profile of cognitive impairment observed in chronic psychosis does not differ from that seen in FEP, suggesting that this cognitive impairment maintains a stable course throughout illness [4]. Because of these findings, cognitive deficits have been proposed as possible endophenotypes of psychotic disorders [5].

The severity of cognitive disturbances in patients with schizophrenia follows a heterogeneous profile [1, 6]. Typically, the magnitude of such impairments ranges from 1 to 2 standard deviations (SD) below the mean of the healthy population, depending on the level of severity and the cognitive domains assessed. When

Leyre Urigüen and Rebeca Diez-Alarcia (eds.), *Schizophrenia: Methods and Protocols*, Methods in Molecular Biology, vol. 2687, https://doi.org/10.1007/978-1-0716-3307-6_10,

alterations are defined as at least 1 SD below the mean of the healthy population, about 80% of patients with schizophrenia present some kind of deficit in at least two cognitive domains when assessed by an exhaustive neuropsychological evaluation [4].

Furthermore, cognitive deficits have been proposed as a major disability determinant and predictor of impaired psychosocial functioning in schizophrenia. Among main cognitive domains affected in schizophrenia, the social cognition domain has been posed as a moderator between neurocognitive performance and real-world functionality [7]. In this regard, research on social cognition deficits has increased over the years, covering the areas of social perception and social behavior, theory of mind, and affective empathy. These abilities have been associated with the capacity to establish and sustain interpersonal relationships, as well as the ability to infer others' intentions and points of view or accurately identify their emotions [7, 8]. Finally, cognitive impairment, particularly deficits in the domain of social cognition, is acknowledged as one of the most significant predictors of functional prognosis in patients with schizophrenia.

As stated earlier in this introduction, assessing the impact of cognitive deficits and their impact on psychosocial functioning is essential to understanding their true impact. Throughout this chapter, we will provide a concise overview of key aspects to consider when assessing neuropsychological performance, as well as the main clinical approaches to address neurocognitive deficits and estimate their impact in overall functionality in schizophrenia spectrum disorders.

2 Materials

2.1 Neuropsychological Batteries and Tests for Cognitive Assessment in Schizophrenia

1. There is currently no standardized neuropsychological protocol for evaluating the presence and severity of cognitive deficits in patients with psychotic disorders.
2. In order to homogenize criteria for the extent of cognitive impairment in schizophrenia in clinical trials, the MATRICS expert (Neurocognition Subcommittee for Measurement and Treatment Research to Improve Cognition in Schizophrenia) initiative proposed the inclusion of seven specific cognitive domains mostly affected in schizophrenia [9]:

   I. Speed of processing
   II. Working memory
   III. Attention/vigilance
   IV. Verbal learning and memory
   V. Visual learning and memory
   VI. Reasoning and problem-solving
   VII. Social cognition

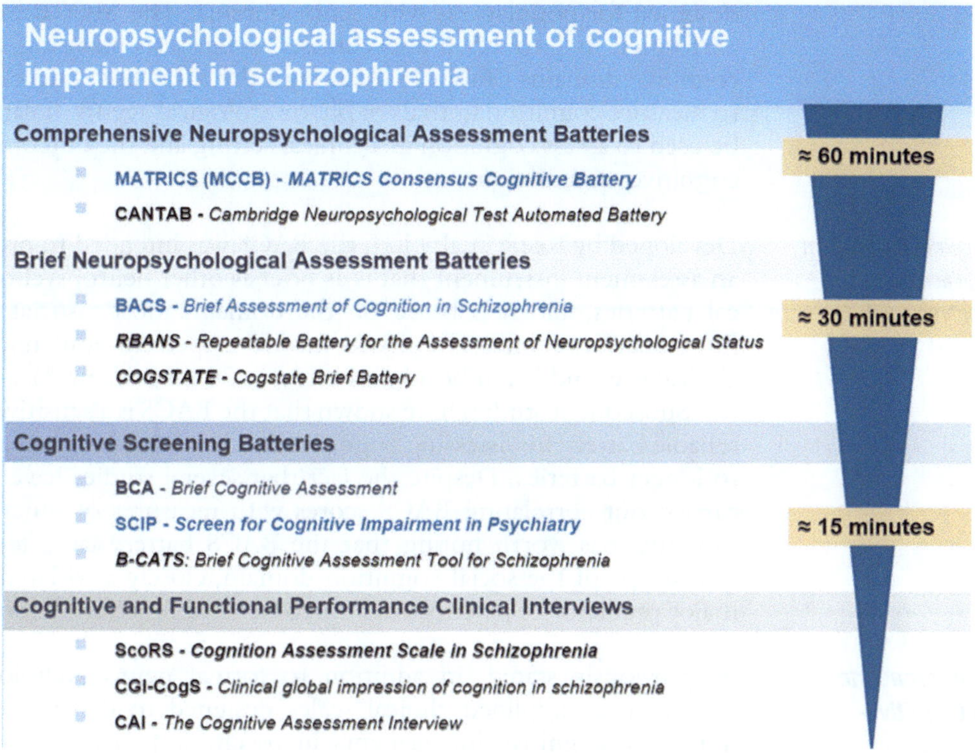

Fig. 1 Main neuropsychological assessment batteries and tests for the assessment of cognitive deficits in clinical and research settings

3. Overall, when assessing a patient with psychosis, it is highly recommended to start with a general assessment to define the patient's average cognitive functioning.

As depicted in Fig. 1, apart from the classical paper-based neuropsychological test, which are not discussed further in this chapter, there is a wide selection of neuropsychological batteries for assessing cognition in schizophrenia. For further information regarding specific neuropsychological testing by a cognitive domain, we recommend consulting the Lezak et al. handbook [10]. Below, we will present two of the most widely used cognitive batteries based on clinical and research practice for the assessment of cognitive deficits in schizophrenia, including the MATRICS Consensus Cognitive Battery (MCCB) and the Brief Assessment of Cognition in Schizophrenia (BACS).

2.1.1 MATRICS Consensus Cognitive Battery (MCCB)

The MATRICS Consensus Cognitive Battery (MCCB) initiative was begun by the National Institute of Mental Health (NIMH) to assist in the development of new therapeutic drugs for schizophrenia that would enhance and improve cognition. Thus, unlike the Wechsler scales or the Repeatable Battery for the Assessment of Neuropsychological Status (RBANS), the MCCB was specifically designed for populations with schizophrenia. The MCCB can be performed in 1 to 1.5 h and consists of ten tests assessing seven cognitive domains (Table 1). Although the MCCB was developed to measure cognition in trials of pharmacological agents, it can also be used to assess cognition in a clinical setting and the outcomes of cognitive remediation.

2.1.2 Brief Assessment of Cognition in Schizophrenia (BACS)

Developed by Keefe et al. [12], the BACS was intended to provide an assessment instrument that is as brief as other neuropsychological batteries, and it is based on the domains most associated to functional outcomes. The BACS was developed for academic and clinical use, and it can be administered in 30 min (Table 2).

Subsequent studies have shown that the BACS is a sensitive and reliable battery for assessing cognitive functioning when compared to longer batteries. Despite the fact that several studies have been carried out correlating BACS scores with measures of functional capacity, it is worth noting that the BACS battery lacks for the assessment of the social cognition domain, closely associated as a major predictor of psychosocial outcomes and functional recovery.

2.2 Approach to Main Cognitive-Functional Performance Assessment Scales

As previously stated, in addition to formal neuropsychological assessments, simplified clinical scales designed to determine the impact of cognitive impairments in psychotic patients could be used. Although showing utility in predicting psychosocial functioning of patients, correlations between the scores obtained in comparison with extensive neuropsychological examinations are limited.

(i) Clinical interviews for the detection of functional alterations—self/hetero-applied by patient, clinician, or relatives.
(ii) Provide a comprehensive picture of the cognitive alterations and their degree of impact on the patient's life.
(iii) Few items: easy to apply.
(iv) Possible inter-rater bias: subjective measurements.
(v) Despite their close relationship with daily functioning, they are not instruments for measuring functional performance per se.

2.2.1 Cognition Assessment Scale in Schizophrenia (ScoRS)

(i) Interview-based instrument with the patient and informant—18 items (application 10–15 min).
(ii) Two separate sources of information are used to generate three different ratings: (a) an interview with the patient, (b) an interview with a patient's informant (the person who has the most regular contact with the patient in everyday situations, such as family members, friends, or a social worker), and (c) a subjective interviewer rating.

Table 1
MATRICS Consensus Cognitive Battery (MCCB)

Cognitive domain	Test	Time of application (minutes)	Order of administration
Speed of processing	Trail Making Test – Part A	2.1	1st
	BACS: Symbol coding subtest	3	2nd
	Category Fluency – Animal naming	2	8th
Attention/vigilance	Continuous Performance Test – Identic Pairs (IP)	13.4	10th
Working memory (verbal and non-verbal)	Non-verbal: Wechsler–III. Spatial Span task	5.1	4th
	Verbal: Letter-Number Span	5.9	5th
Visual learning	Brief Visuospatial Memory Test - Revised	4.7	7th
Verbal learning	Hopkins Verbal Learning Test- Revised	5	3rd
Reasoning and problem solving	Neuropsychological Assessment Battery – NAB: Mazes	11.2	6th
Social cognition	Mayer-Salovey-Caruso Emotional Intelligence Test – Managing Emotions	12	9th

Table 2
Brief Assessment of Cognition in Schizophrenia (BACS)

Cognitive functions assessed	Test	Time of application (seconds)	Order of administration
Verbal learning and memory	Verbal Learning	-	1st
Working memory	Digit Sequencing	-	2nd
Motor function and speed	Token Motor	60 s	3rd
Verbal fluency	Semantic verbal fluency	60 s	4rd
	Phonemic verbal fluency	60s per letter	
Attention and processing speed	Symbol Coding	90 s	5th
Executive function	Tower of London	20s /item (max 22 items)	6th

(iii) Clinician provides an estimate global rating score based on his or her clinical observations of the patient.

(iv) Allow clinicians to assess neurocognitive deficits and the extent to which they affect the patient's daily functioning (as reported by patients and their relatives or primary caregivers).

(v) Assesses patients' ability to perform daily tasks in relation to cognitive level and functionality.

2.2.2 Cognitive Assessment Interview (CAI)

(i) Interview-based instrument comprising ten items. Developed from the CGI-CogS [11] and the SCoRS [12].

(ii) Based on the seven cognitive domains proposed by the MATRICS consortium.

(iii) The CAI incorporates all available sources of information and assigns a score based on patient's neurocognitive state and how neurocognition affects daily functioning.

(iv) For CAI ratings, it is specifically important that the interviewers attempt to make ratings based on impressions about the patient's cognitive functioning without undue influence from severity/change of other symptomatology of the disorder.

(v) The score is calculated by rating the patient interview, the informant interview, and the overall severity of cognitive impairment separately.

(vi) When an informant is not available, the CAI may be preferred over the ScoRS as patient-only assessment demonstrates better test-retest reliability.

2.3 Measurement of Day-to-Day Functionality: The Role of Functional Performance Scales

1. Tasks based on patient performance in the real world.
2. Increasing relevance. For approval of cognitive enhancers, FDA requires:
   (a) Evidence of improvements in neurocognitive tests.
   (b) Relevance of such improvements to everyday functioning (ecological validity).

As previously stated, the strong link between cognition and functional outcomes supports the need to evaluate functional testing in order to better understand patient outcomes. In this regard, improvement in a cognitive performance outcome, as indicated by the FDA-NIMH-MATRICS Guidelines for Clinical Trial Design for Cognitive-Enhancing Drugs, would be a necessary but not sufficient requirement for novel treatment approval. In addition to the requisite changes in cognitive performance, improvement on a co-primary functional measure with more ecological validity measures will be required [13].

2.3.1 Global Measures of Functioning in Community

These scales evaluate the overall functioning of the subject:

1. *Global Activity Assessment Scale (GAF)*
   (i) Clinically used to subjectively assess the social and occupational functioning and psychological symptoms of adults.
   (ii) It measures how much a person's symptoms affect their day-to-day life on a scale of 0 to 100.
   (iii) Limitation: inability to differentiate psychiatric symptoms from relational, social, and occupational functioning.
2. *Social and Occupational Functioning Scale (SOFAS)*
   Valid clinician-rated scale used to assess social and occupational functioning based on medical conditions.
   (i) Developed in the context of need for a specific assessment tool of global functioning, not directly affected by psychopathology.
   (ii) Include impairments in functioning due to physical limitations, as well as those due to mental impairments.

(iii) The scale is based on a continuum of functioning, from 0 to 100, with higher scores indicating better functioning.

(iv) The final score is obtained based on three dimensions: level of autonomy, social functioning, and occupational functioning.

(v) Main differences from the GAF scale: It focuses exclusively on the individual's level of social and occupational functioning and is not directly influenced by the overall severity of the individual's psychological symptoms.

(vi) Any impairment in social and occupational functioning that is due to general medical conditions is considered in making the SOFAS rating.

2.3.2 Measurement of the Subject's Functional Performance: Functional Capacity

UCSD Performance-Based Skills Assessment (UPSA)

(a) Valid and reliable for measuring changes in functional capacity.

(b) Distinguishes between different levels of disease severity and between the functioning of patients and healthy subjects.

(c) Assesses six domains of basic daily living skills:

Household duties (cooking, grocery, shopping).

Communication (phone calls).

Public transport use.

Finances (counting the change and settling bills).

Recreational activity planning.

Medication management.

(d) Through role-play and use of props recreates everyday situations (role-play).

(e) Examiner will award scores from 0 to 100. High scores improve functioning.

(f) Dual purpose: (a) to measure the ability to function autonomously and (b) to test the effectiveness of interventions on functionality.

(g) Limitations: designed for older patients. Young patients possible ceiling effect.

(h) There is available a brief-version of the UPSA (UPSA-B), which can be administered in 10–15 min including the communication and finance subscales of the UPSA.

The Brief International Functional Capacity Assessment (BIFCA)

(i) Intended to provide evidence of level of functional capacity for everyday functioning in patients with schizophrenia. Time of application of the scale is 15–20 min.

(ii) Has been validated as multicultural measure of functional capacity that incorporates main aspects of the UPSA scale.

(iii) Developed to be used as a co-primary outcome measure in clinical trials of cognitive-enhancing drugs for adults with schizophrenia spectrum disorders.

(iv) Contains three modules designed to evaluate functional capacity in key areas of basic living skills (the "comprehension and planning," "household management," and "work and productivity" domains).

(v) Individuals demonstrate their ability to plan, identify problems and generate solutions, and persist in an activity through standardized performance-based simulations of daily activities.

3 Methods

3.1 Prior Considerations for the Neuropsychological Assessment of Neurocognitive Deficits

When evaluating main cognitive dimensions affected in people with psychotic disorders, it is critical to consider which factors may influence their performance:

1. Whenever possible, it is recommended to use extensive neuropsychological assessments in order to determine which areas are most affected and the degree of impairment of each patient. Indeed, such tests are especially useful in the early stages of psychotic disorders. Early detection and treatment of cognitive disturbances may reduce the impact on patients' psychosocial functioning (social, occupational, and autonomy areas).
2. Reliable and accurate information on patients' actual cognitive status requires accurate and practical assessment tools. When it is not possible to administer comprehensive neuropsychological batteries, it is recommended to use equivalent validated short batteries and tests.
3. The use of screening batteries for global cognitive status, such as the Mini-Mental State Examination (MMSE) or the Montreal Cognitive Assessment (MoCA), which are widely used in other medical specialties, is not recommended. Specific screening batteries for patients with psychotic disorders have been developed as part of new initiatives, with the Screen for Cognitive Impairment in Psychiatry (SCIP) brief instrument standing out. These brief batteries have the advantage of being composed of fewer cognitive tests, which results in a shorter application time and allows for a brief assessment of the overall cognitive dimensions in psychiatric populations.

3.1.1 How and When to Assess Cognition in Psychotic Disorders

1. Ensure patient clinical symptomatology stability prior to cognitive assessment. If not achieved, wait until symptom remission (*see* **Note 1**).
2. Whenever possible (considering available resources and time), it is recommendable to perform an exhaustive neuropsychological examination (*see* **Note 2**).
3. Use. . .chose. . .reliable and practical assessment instruments in order to assess real cognitive state.

4. Neuropsychological assessment should be performed by qualified professionals with specific training in the field neuropsychological testing and the interpreting of the results.

3.2 Neuropsychological assessment

3.2.1 Before Examination

Some prior considerations should be taken into account before initiating the neuropsychological assessment:

1. The patients should be tested by a trained specialist (*see* **Note 3**).
2. The time duration can vary based on the referral question, the patient's symptoms, and the examiner expertise.
3. Revise medical history: Consider if there are any sensory-perceptual deficits that could interact in patient's performance (*see* **Note 4**).
4. What do we want to measure? There is no unique protocol, wide diversity of neuropsychological batteries, and tests depending on specific demands.
5. Structure and organize neuropsychological examination: Assessment must be sequential (*see* **Note 5**).
6. Choose practical tests taking into account the means and context (time, location, training, etc.) (*see* **Note 6**).
7. Challenges inherent in neuropsychological tests: finding pure measures of a single domain. It is challenging to assess a given cognitive process on its own since the proper execution of a task usually requires the integration of multiple cognitive processes.
8. Other potential issues: lack of time, resources or space, and funding and professionals with specific training.
9. Assessment should consider a biopsychosocial perspective including all available information (*see* **Note** 7).

3.2.2 During Assessment

1. Setting: The room must have adequate conditions (quiet environment, adequate room temperature, comfortable furnishings).
2. Check that you have all the material before examination begins (clinical scales, neuropsychological testing material, office supplies, and a timekeeper) (*see* **Note 8**).
3. Explain the patient what to expect from a neuropsychological assessment. Resolve any questions raised during the initial interview.
4. Obtain information regarding patient's sociodemographic, medical, and psychiatric clinical data (*see* **Note 9**).
5. Consider major factors that could potentially influence cognitive performance:

(i) Age and sex.
(ii) Education level and premorbid intelligence.
(iii) Hand dominance.
(iv) Fatigue and physiological states (hunger, sleep patterns, etc.).
(v) Substance abuse influence.
(vi) Nicotine and caffeine intake.
(vii) History of neurological disease.
(viii) Practice effect.
(ix) Clinical status: acute *vs.* remission phase.
(x) Motivation, interest, and collaboration rate.
(xi) Comprehension difficulties or language barriers.
(xii) Impact of concomitant psychopharmacological treatments (i.e., benzodiazepines, anticholinergic burden).

3.2.3 Major Considerations to Be Taken After a Neuropsychological Assessment

1. A key consideration in interpreting the results is to take into account the subject's premorbid functioning and trajectory in relation to their sociocultural background.
2. Individual test scores are interpreted by comparing them to normative data (considering age, education, gender) and expected levels of functioning.
3. Most importantly, combine all available data sources to estimate the real cognitive functioning of the patient (test results, informants, and clinical judgment).
4. Relevance in early detection of strengths and weaknesses. Planning of personalized therapeutic interventions.
5. Cognitive assessment, based on validated and reliable instruments, should be considered for all patients with schizophrenia, both at baseline and throughout long-term treatment.
6. Prognostic significance of cognitive deficits in the clinical diagnosis and daily patient functioning.

4 Notes

1. Cognitive examination during acute phase should be avoided as it may interfere with the detection of actual cognitive status and the assessment of individual cognitive weaknesses and strengths.
2. It is advisable to delimit affected areas and the degree of impairment in order to find strengths and weakness and plan the most appropriate intervention in each case.

3. Neuropsychological examination should be performed by an expert on neuropsychological testing, which has been trained in how to administer, to score, and interpret the obtained information.
4. Revise medical history to assess if the patient presents any sensory-perceptual deficits. Advise the patient in advance to bring hearing aids or glasses if necessary.
5. If longer neuropsychological assessment is required, consider divide in separated sessions (2 days).
6. Consider the target population to ensure tolerability: brief *vs.* more extensive neuropsychological batteries.
7. Clinical anamnesis must include the etiology, course of the disease, personal and family medical history, and psychosocial and relevant academic information.
8. Office supplies, like a pen, a pencil, and sheets to reflect annotations during the assessment. It is advisable to have a timekeeper in case there are tests that consider the time elapsed during a task.
9. In order to assist in interpretation of neuropsychological data, all available sources should be taken into consideration (neuropsychological performance).

References

1. Mohamed S, Paulsen JS, O'Leary D, Arndt S, Andreasen N (1999) Generalized cognitive deficits in schizophrenia: a study of first-episode patients. Arch Gen Psychiatry 56: 749–754
2. Bowie CR, Harvey PD (2006) Cognitive deficits and functional outcome in schizophrenia. Neuropsychiatr Dis Treat 2:531–536
3. Snitz BE, Macdonald AW 3rd, Carter CS (2006) Cognitive deficits in unaffected first-degree relatives of schizophrenia patients: a meta-analytic review of putative endophenotypes. Schizophr Bull 32:179–194
4. Segarra R, Zabala A, Eguíluz J, Ojeda N, Elizagarate E, Sánchez P et al (2011) Cognitive performance and smoking in first-episode psychosis: the self-medication hypothesis. Eur Arch Psychiatry Clin Neurosci 261:241–250
5. Yeo RA, Gangestad SW, Walton E, Ehrlich S, Pommy J, Turner JA et al (2014) Genetic influences on cognitive endophenotypes in schizophrenia. Schizophr Res 156:71–75
6. Fioravanti M, Bianchi V, Cinti ME (2012) Cognitive deficits in schizophrenia: an updated meta-analysis of the scientific evidence. BMC Psychiatry 12:64
7. Green MF, Horan WP, Lee J (2019) Nonsocial and social cognition in schizophrenia: current evidence and future directions. World Psychiatry 18:146–161
8. Henry JD, von Hippel W, Molenberghs P, Lee T, Sachdev PS (2016) Clinical assessment of social cognitive function in neurological disorders. Nat Rev Neurol 12:28–39
9. Nuechterlein KH, Barch DM, Gold JM, Goldberg TE, Green MF, Heaton RK (2004) Identification of separable cognitive factors in schizophrenia. Schizophr Res 72:29–39
10. Lezak MD, Howieson DB, Loring DW, Fischer JS (2004) Neuropsychological assessment. Oxford University Press, New York
11. Bilder R, Ventura J, Cienfuegos A (2003) Clinical Global Impression of Cognition in Schizophrenia (CGI-CogS) manual and rating sheet. UCLA Department of Psychiatry, Los Angeles
12. Keefe R, Poe M, Walker T, Kang J, Harvey P (2006) The schizophrenia cognition rating scale: an interview-based assessment and its relationship to cognition, real-world functioning, and functional capacity. Am J Psychiatry 163:426–432

13. Buchanan RW, Keefe RS, Umbricht D, Green MF, Laughren T, Marder SR (2011) The FDA-NIMH-MATRICS guidelines for clinical trial design of cognitive-enhancing drugs: what do we know 5 years later? Schizophr Bull 37: 1209–1217

Chapter 11

Clinical Management and Assessment of the Early Stages of Psychosis: A Multidimensional Approach

Oihane Mentxaka, María Recio-Barbero, Janire Cabezas-Garduño, and Rafael Segarra

Abstract

Early intervention in psychosis has emerged as an integrated clinical and research strategy for the comprehensive care of people presenting with a first psychotic episode (FEP). In this sense, a multidisciplinary approach is essential for early detection and assessment of preliminary stages of the illness. The main objectives of early intervention programs include the implementation of specialized interventions aimed at promoting functional recovery and improving quality of life. In this chapter, we will describe the main advantages of specialized early intervention programs in psychosis. We will also describe the main aspects of assessment that need to be considered when approaching an FEP.

Key words First-episode psychosis, Schizophrenia, Early intervention program, Assessment, Intervention

1 Introduction

A first psychotic episode (FEP) occurs when a broad range of psychotic symptoms, including hallucinations, delusions, and/or disorganized thinking, occur for the first time in an individual. It can correspond to the onset of a number of serious mental disorders, such as schizophrenia, schizophreniform disorder, acute psychotic disorder, delusional disorder, bipolar disorder, major depressive disorder with psychotic symptoms, and others. The onset of FEP is therefore a particularly complex and stressful event, affecting both the individual and their immediate environment, and compromising the potential development of the person suffering from it [1].

Before the onset of the disease, a number of subtle early signs have been described. Disturbances in perception and beliefs, persistent unusual thoughts that cannot be put aside regardless of what

Leyre Urigüen and Rebeca Diez-Alarcia (eds.), *Schizophrenia: Methods and Protocols*, Methods in Molecular Biology, vol. 2687, https://doi.org/10.1007/978-1-0716-3307-6_11,

others believe, difficulties with concentration and attention, and a sudden decrease in self-care have all been described [1, 2].

FEP often occurs during adolescence, typically between the ages of 15 and 35, a critical time in life for personal, educational, and professional development [3, 4]. In addition, a second peak of onset has been reported in women between the ages of 40 and 55 [4, 5]. In this context, early intervention in psychosis aims to minimize the social and economic impact of disability caused by the illness and to achieve functional recovery [6, 7].

In this sense, knowing when to seek help and how to access specific mental health resources is a major dilemma for some families and patients. Furthermore, there is often a significant delay between the onset of symptoms and the subsequent diagnosis of FEP, leading to delays in treatment and consequent adverse effects on functional outcomes [8].

There is international consensus on the importance of early intervention in the stages of FEP, which includes improving the course of the disease by moving beyond response to treatment to functional recovery and improved quality of life [9]. The aim is to promote the early inclusion of people with FEP in work or study and to achieve a level of satisfaction with these activities equivalent to that of the population of the same age and sex.

In order to reduce the burden of the disease, specialized units for the treatment of first psychotic episodes have been created on the basis of early intervention programs (EIPs). The aim of these specialized units is to detect early symptoms of the illness and to implement a treatment plan (including psychotherapy and pharmacological interventions) once the onset of the illness is confirmed [9].

There is agreement on the benefits of implementing EIPs in the clinical management of FEPs. Main benefits of these programs include:

- As early intervention takes place after the onset of the psychotic illness, it has been consistently shown to prevent psychosocial deterioration by reducing psychosocial disability and as a driver of functional recovery.
- Early intervention has also shown benefits in the first years after onset. Follow-up and treatment during the first 5 years of illness are recommended. The ultimate goal is to minimize the long-term psychosocial and functional impact and reduce the burden of the disease to prevent chronic stages.

Early intervention during the first 5 years after debut (the so-called critical period) has been identified as an important critical period to prevent relapse and residual symptoms. Considering that a high percentage of FEP patients, up to 80%, will relapse within the first 5 years after debut, early intervention may contribute to identify, intervene, and improve the major clinical and psychosocial impact and long-term prognosis [10, 11]. Therefore, the development of multidisciplinary and translational units, based on clinical practice and research, is crucial for the advancement of scientific knowledge and the further development and implementation of novel strategies in the early stages of psychosis. Early intervention in psychosis health services could be accessible to all people presenting with a FEP, regardless of age or duration of untreated psychosis. Within this framework, several programs and units for early intervention in psychosis have been developed in our country. Below we outline the main features and the design of the intervention program of the Early Psychotic Episodes Unit—CRUPEP intervention program developed at the Cruces University Hospital (Bilbao, Spain).

A comprehensive and individualized assessment of the patient's situation should be carried out in order to plan personalized intervention strategies to improve the quality of life, functioning, and prognosis of people affected by a PEP in the short and medium term.

Clinical Assessment

The diagnosis and management of the early stages of psychosis can present clinicians with a number of challenges. In this context, the onset of illness might encompass a wide range of symptoms, making clinical management and diagnosis difficult.

Potential referral sources for the program include (Fig. 1):

1. Psychiatrists working in the emergency department or acute hospitalization unit of the University Hospital of Cruces
2. Psychiatrists from the local mental health services
3. Other health professionals in our area, such as family doctors, pediatricians, and neurologists, among others

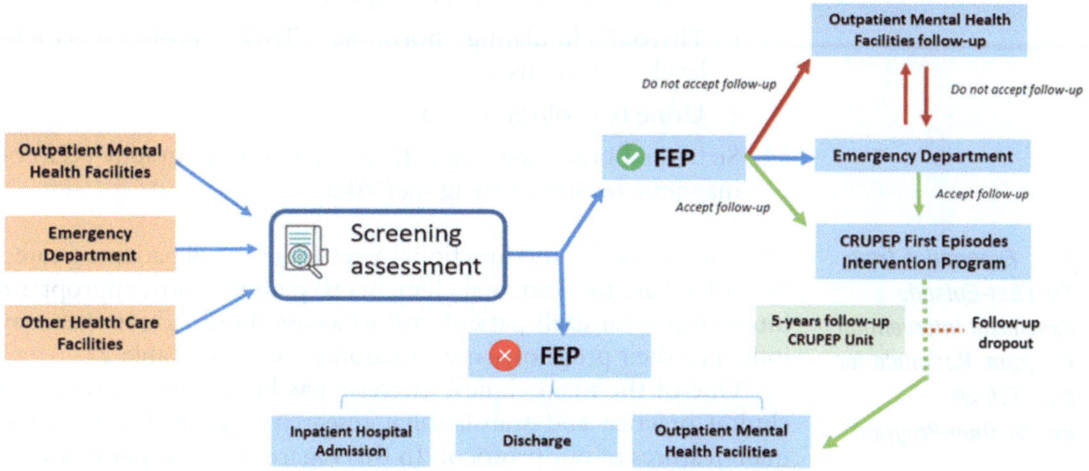

Fig. 1 CRUPEP first episode psychosis intervention program referral scheme

Table 1
Epidemiology of first-episode psychosis and clinical, therapeutic, and prognostic considerations

First-episode psychosis spectrum			
15% of FEPs	Major depressive disorder with psychotic symptoms Bipolar disorder with psychotic symptoms Acute intoxication Cycloid psychosis (*Bouffée délirante, reactive psychosis, etc.*)	Schizotypal disorder Brief psychotic disorder Schizophreniform disorder Schizophrenia Delusional disorder Induced psychotic episode	85% of FEPs

Adapted from McGorry et al. [12]

One of the main aims of our early psychotic intervention program is to incorporate translational research into clinical practice. Figure 1 shows the main features of our protocol. In this context, the assessment procedures and timelines are described below.

Once the referral form is received, the patient will normally be assessed within 72 h.

1. Clinical assessment and differential diagnosis: differential diagnosis with other clinical manifestations including substance misuse (Table 1).
2. Physical examination: including cardiovascular or metabolic disorders and related lifestyle factors, such as smoking status, diet, and exercise.
3. Laboratory testing.
   - Complete blood count
   - Comprehensive metabolic panel, including serum electrolytes (to rule out metabolic disorders), blood glucose, and liver and kidney panel tests
   - Serum vitamin B12 and folate levels
   - Thyroid-stimulating hormone (TSH), prolactin (PRL) levels, and cortisol
   - Urine toxicology screen
4. Structural brain imaging with computed tomography (CT) or magnetic resonance imaging (MRI).

1.1 Experience from the First-Episode Psychosis Intervention Program: Rationale for the CRUPEP Intervention Program

The clinical, social, and functional assessments of all patients treated by an FEP are fundamental elements to plan the most appropriate intervention for each patient and to assess those factors that may influence their prognosis and functional recovery (Table 2).

One of the main clinical interests has been the integration of clinical practice and translational research. Figure 2 depicts the main features of our protocol. In this regard, the evaluation procedures and timelines are outlined below:

Table 2
Main clinical assessment scales and time points

	Baseline	6-month follow-up	1-year follow-up	2-year follow-up	3-year follow-up	4-year follow-up	5-year follow-up
Sociodemographic data	x						
Medical assessment	x						
Prior personal and family history	x						
Clinical psychiatric diagnosis	x	x	x	x	x	x	x
Drug use history	x	x	x	x	x	x	x
Psychopharmacological treatment	x	x	x	x	x	x	x
Laboratory testing—routine blood tests and toxicology	x	x	x	x	x	x	x
Relevant clinical data—psychotic relapses, inpatient acute admissions and psychiatric emergency department consultations	x	x	x	x	x	x	x
Clinical assessment							
Treatment adherence—Morisky-Green scale (DAI)		x	x	x	x	x	x
Positive and Negative Syndrome Scale (PANSS)	x	x	x	x	x	x	x
Brief Negative Symptom Scale (BNSS)	x	x	x	x	x	x	x
Montgomery-Åsberg Depression Rating Scale (MADRS)	x	x	x	x	x	x	x
Young Mania Rating Scale (YMRS)	x	x	x	x	x	x	x
Clinical Global Impression Scale (CGI)	x	x	x	x	x	x	x
Cannon-Spoor's Premorbid Adjustment Scale (PAS)	x						
Strauss and Carpenter Prognostic Scale (SCPS)	x						
Functioning Assessment Short Test (FAST)	x	x	x	x	x	x	x
Zarit Burden Interview—for primary caregivers	x						
Childhood Trauma Questionnaire	x						
UKU Side Effect Rating Scale (UKU)		x					
Neurocognitive assessment	x		x	x	x	x	x

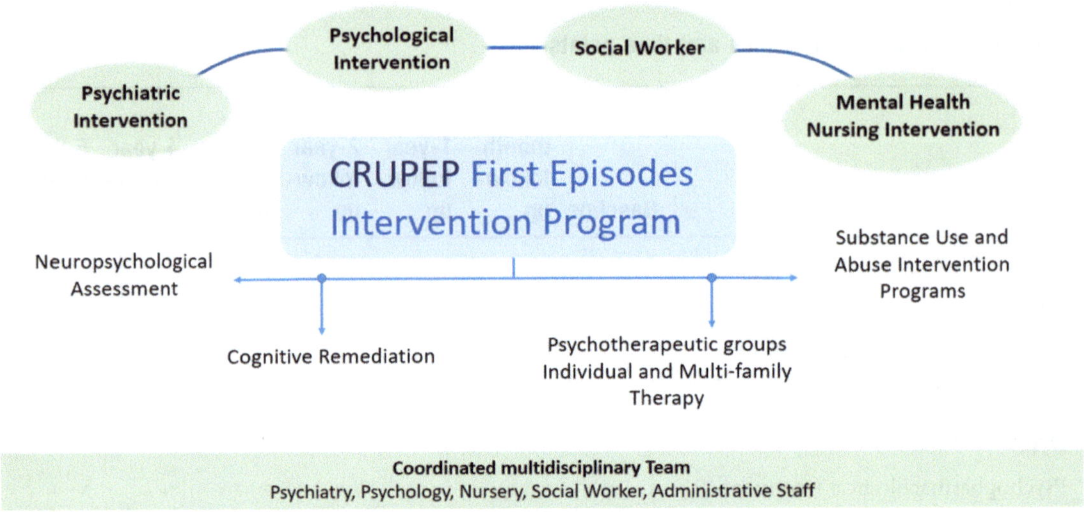

Fig. 2 Clinical rationale for the CRUPEP first episode intervention program

1. A trained psychiatrist from the early psychosis intervention program—CRUPEP—conducts the initial interview (baseline interview). The diagnostic protocol is followed, and once a primary diagnosis of psychosis has been established, psychiatric consultations are flexible, depending on need. They can be weekly or monthly, depending on the patient's needs and clinical progress. It is important to involve family members and/or carers.
2. Coordination with nursing staff is essential as part of the psychoeducational intervention and approach.
3. A neuropsychological evaluation of the patient's cognitive state is performed as part of the therapeutic approach. This evaluation is carried out in stages (once a year during the 5 years of follow-up). Similarly, the individual relevance of different psychotherapeutic approaches (individual, group, or family) is assessed in collaboration with the rest of the team.

2 Materials

2.1 Initial Multidisciplinary Assessment

2.1.1 Clinical Assessment

Includes a comprehensive clinical and social assessment, including evaluation of the patient's premorbid adjustment, cognitive and functional performance, illness awareness, and physical health status. Initial assessments focus on establishing a timeline of symptoms based on a detailed history and physical examination:

1. Assessment of premorbid and prodromal symptoms
2. Developmental milestones addressing social, cognitive, and motor developmental skills along the developmental history of childhood and adolescence

3. Prior diagnosis, treatment, and follow-up on mental healthcare facilities
4. Family history of psychiatric disorders
5. History of substance use and abuse
6. Physical health and well-being, including healthy habits, weight, smoking status, nutritional habits, and physical activity
7. Psychological and psychosocial development, including social support networks and relationship with peers. Assessment of experimentation of previous adverse childhood events or trauma-related experiences
8. Sociocultural determinants, including culture, ethnicity, and socio-economic status, among others.
9. Occupational and educational determinants, including school attendance, educational attainment and employment, and activities of daily living

A. *Therapeutic options*: include psychotropic medication combined with psychotherapeutic interventions based on the clinical needs of each patient. A number of considerations need to be taken into account when considering psychopharmacological interventions.

Based on National Institute for Health and Care Excellence (NICE) guidelines [13], the choice of antipsychotic medication should be made jointly by the patient and the clinician. It is essential to consider both the potential benefits and potential side effects of each drug, including:

(i) Metabolic (including weight gain and diabetes)
(ii) Extrapyramidal (including akathisia, dyskinesia, and dystonia)
(iii) Cardiovascular (including prolonging the QT interval)
(iv) Hormonal (including increasing plasma prolactin)
(v) Other relevant aspects (including unpleasant subjective experiences)

2.1.2 Family and Social Assessment

Includes assessment of social support network, family, and general social situation. The family assessment is carried out in parallel with the clinical assessment of the patient. A family assessment is conducted with the patient to better understand the structure and functioning of the family. The main objectives are as follows:

(i) Identify the family member or primary caregiver (the person who spends most time with the patient and is responsible for his/her care).
(ii) Assess key needs and identify the family's skills and abilities.

2.1.3 Psychological Intervention

1. The clinical psychologist conducts focal interventions focused on the management of specific objectives and limited time, when deemed appropriate by the multi-professional team.
2. Individual and/or group psychoeducational therapies, such as family interventions and behavioral activation, are also used to provide management skills for coping with anxiety symptoms and stressful situations.
3. Most evidence-based psychotherapies, such as cognitive behavioral therapy (CBT), psychoeducation, and family intervention groups, are recommended by current evidence. In this regard, CBT enables patients to assess and understand the symptoms of their illness; psychoeducation for patients includes information about the illness, treatment options, social skills, relapse prevention, basic life skills, and problem-solving strategies. Furthermore, the development of effective symptom management strategies is encouraged.
   (i) The aim is to reduce the patient's vulnerability to stressful situations by increasing his or her resources for dealing with conflicts, problems, and interpersonal or biographical tensions, thereby facilitating recovery processes and strengthening adaptation and family, social, educational, and occupational functioning.
   (ii) Interventions will focus on symptom reduction, relapse prevention, social skill acquisition, and social and occupational functioning.
   (iii) They should be tailored to the patient's stage of the disease. Psychological interventions that are currently supported by scientific evidence should be considered in the development of the treatment plan.

2.1.4 Occupational and Employment Assessment

(i) Based on classic vocational rehabilitation programs, the main objective is to provide FEP patients with guidance and counseling to acquire skills for workplace integration and competitive employment.

(ii) The shift from the traditional train-then-place model of vocational rehabilitation (VR) to supported employment programs such as Individual Placement and Support (IPS) represents a shift in the services available to people with neuropsychiatric disorders.

(iii) The impact of IPS on occupational performance is well documented. However, numerous occupational challenges remain for this group, including difficulties with job performance and unwanted job abandonment.

3 Clinical Follow-Up Assessments

1. Integrate clinical assessments into the routine follow-up of the patient, so that a complete assessment of the patient and their environment is carried out throughout the 5 years of follow-up.
2. Implement early detection of "red flag" signs—warning symptoms of relapse.
3. It is essential to monitor all potentially adverse effects of psychotropic medication.
4. Assessment of physical health: metabolic syndrome and healthy habits.
5. Regular clinical re-evaluation allows early identification of those patients who require more intensive or specialized interventions: incomplete clinical remission and functional recovery, high risk of treatment withdrawal, coexisting dual pathology, depressive symptomatology, suicidal ideation, and major social or family problems.
6. Select appropriate clinical tools for a thorough assessment, preferably ones that have been widely used in scientific evidence-based early psychosis programs and validated in the native language of the subject.

References

1. Frangou S, Byrne P (2000) How to manage the first episode of schizophrenia. BMJ 321(7260):522–523
2. Maj M, van Os J, De Hert M, Gaebel W, Galderisi S, Green MF, Guloksuz S, Harvey PD, Jones PB, Malaspina D, McGorry P, Miettunen J, Murray RM, Nuechterlein KH, Peralta V, Thornicroft G, van Winkel R, Ventura J (2021) The clinical characterization of the patient with primary psychosis aimed at personalization of management. World Psychiatry 20(1):4–33
3. O'Donoghue B, Lyne J, Madigan K, Lane A, Turner N, O'Callaghan E, Clarke M (2015) Environmental factors and the age at onset in first episode psychosis. Schizophr Res 168(1–2):106–112
4. Faraone SV, Chen WJ, Goldstein JM, Tsuang MT (1994) Gender differences in age at onset of schizophrenia. Br J Psychiatry 164(5):625–629
5. Eranti SV, MacCabe JH, Bundy H, Murray RM (2013) Gender difference in age at onset of schizophrenia: a meta-analysis. Psychol Med 43(1):155–167
6. Charlson FJ, Ferrari AJ, Santomauro DF, Diminic S, Stockings E, Scott JG, McGrath JJ, Whiteford HA (2018) Global epidemiology and burden of schizophrenia: findings from the global burden of disease study 2016. Schizophr Bull 44(6):1195–1203
7. Lieberman JA, Small SA, Girgis RR (2019) Early detection and preventive intervention in schizophrenia: from fantasy to reality. Am J Psychiatry 176(10):794–810
8. Norman RM, Malla AK, Verdi MB, Hassall LD, Fazekas C (2004) Understanding delay in treatment for first-episode psychosis. Psychol Med 34(2):255–266. https://doi.org/10.1017/s0033291703001119
9. McGorry PD, Killackey E, Yung A (2008) Early intervention in psychosis: concepts, evidence and future directions. World Psychiatry 7(3):148–156. https://doi.org/10.1002/j.2051-5545.2008.tb00182.x. PMID: 18836582; PMCID: PMC2559918
10. Brown E, Bedi G, McGorry P, O'Donoghue B (2020) Rates and predictors of relapse in first-episode psychosis: an Australian cohort study. Schizophr Bullet Open 1(1):sgaa017

11. Murphy B, Brewer W (2011) Early intervention in psychosis: strengths and limitations of services. Adv Psychiatr Treat 17(6):401–407. https://doi.org/10.1192/apt.bp.110.008573
12. McGorry PD, Hartmann JA, Spooner R, Nelson B (2018) Beyond the "at risk mental state" concept: transitioning to transdiagnostic psychiatry. World Psychiatry 17(2):133–142. https://doi.org/10.1002/wps.20514. PMID: 29856558; PMCID: PMC5980504
13. Overview: Psychosis and schizophrenia in adults: Prevention and Management: Guidance [Internet]. NICE. [cited 2022 Dec 10]. Available from: https://www.nice.org.uk/Guidance/CG178

Index

Leyre Urigüen and Rebeca Diez-Alarcia (eds.), *Schizophrenia: Methods and Protocols*, Methods in Molecular Biology, vol. 2687, https://doi.org/10.1007/978-1-0716-3307-6,

N

O

P

Q

R

S

MIX
Papier aus verantwortungsvollen Quellen
Paper from responsible sources
FSC® C105338

If you have any concerns about our products,
you can contact us on
ProductSafety@springernature.com

In case Publisher is established outside the EU,
the EU authorized representative is:
Springer Nature Customer Service Center GmbH
Europaplatz 3, 69115 Heidelberg, Germany

Printed by Libri Plureos GmbH
in Hamburg, Germany